The Andean Cloud Forest

Randall W. Myster
Editor

The Andean Cloud Forest

Springer

Editor
Randall W. Myster
Department of Biology
Oklahoma State University
Oklahoma City, OK, USA

ISBN 978-3-030-57343-0   ISBN 978-3-030-57344-7   (eBook)
https://doi.org/10.1007/978-3-030-57344-7

© Springer Nature Switzerland AG 2021, Corrected Publication 2021
This work is subject to copyright. All rights are reserved by the Publisher, whether the whole or part of the material is concerned, specifically the rights of translation, reprinting, reuse of illustrations, recitation, broadcasting, reproduction on microfilms or in any other physical way, and transmission or information storage and retrieval, electronic adaptation, computer software, or by similar or dissimilar methodology now known or hereafter developed.
The use of general descriptive names, registered names, trademarks, service marks, etc. in this publication does not imply, even in the absence of a specific statement, that such names are exempt from the relevant protective laws and regulations and therefore free for general use.
The publisher, the authors, and the editors are safe to assume that the advice and information in this book are believed to be true and accurate at the date of publication. Neither the publisher nor the authors or the editors give a warranty, expressed or implied, with respect to the material contained herein or for any errors or omissions that may have been made. The publisher remains neutral with regard to jurisdictional claims in published maps and institutional affiliations.

This Springer imprint is published by the registered company Springer Nature Switzerland AG
The registered company address is: Gewerbestrasse 11, 6330 Cham, Switzerland

Dedicated to the memory of my Norwegian grandmother, Dora.

"Det som er elsket er alltid vakkert".

Contents

1 Introduction 1
Randall W. Myster

**2 Dynamics of Andean Treeline Ecotones: Between Cloud
Forest and Páramo Geocritical Tropes** 25
Fausto O. Sarmiento

**3 Análisis Regional En Ecosistemas De Montaña En Colombia:
Una mirada desde la funcionalidad del paisaje y los servicios
ecosistémicos** 43
Paola Isaacs-Cubides, Julián Díaz, and Tobias Leyva-Pinto

4 Ecohydrology of Tropical Andean Cloud Forests 61
Conrado Tobón

**5 Litterfall in Andean Forests: Quantity, Composition, and
Environmental Drivers** 89
Wolfgang Wilcke

**6 Arbuscular Mycorrhizal Fungi and Ectomycorrhizas in the
Andean Cloud Forest of South Ecuador** 111
Ingeborg Haug, Sabrina Setaro, and Juan Pablo Suárez

**7 Nesting Ecology of the Tucuman Amazon (*Amazona tucumana*)
in the Cloud Forest of Northwestern Argentina** 131
Luis Rivera and Natalia Politi

8 Adaptive Strategies of Frugivore Bats to Andean Cloud Forests 151
Adriana Ruiz and Pascual J. Soriano

**9 Neotropical Biodiversity: Hypotheses of Species Diversification
and Dispersal in the Andean Mountain Forests** 177
Angela M. Mendoza-Henao and Juan C. Garcia-R

viii

10 Mapping Hydrological Ecosystem Services and Impacts of Scenarios for Deforestation and Conservation of Lowland, Montane and Cloud-Affected Forests 189
Mark Mulligan

11 Conclusions, Synthesis, and Future Directions 219
Randall W. Myster

Correction to: Análisis Regional En Ecosistemas De Montaña En Colombia: Una mirada desde la funcionalidad del paisaje y los servicios ecosistémicos C1

Contributors

Julián Díaz Instituto de Investigación de Recursos Biológicos Alexander von Humboldt, Bogota, Colombia

Juan C. Garcia-R Molecular Epidemiology and Public Health Laboratory, Hopkirk Research Institute, School of Veterinary Science, Massey University, Palmerston North, New Zealand

Ingeborg Haug Universitat Tubingen, Tubingen, Germany

Paola Isaacs-Cubides Instituto de Investigación de Recursos Biológicos Alexander von Humboldt, Bogota, Colombia

Tobias Leyva-Pinto Universidad Nacional de Colombia, Bogota, Colombia

Conrado Tobon Universidad Nacional de Colombia, Medellin, Colombia

Angela M. Mendoza-Henao Departamento de Zoología, Instituto de Biología, Universidad Nacional Autónoma de México, Mexico City, México

Mark Mulligan Department of Geography, King's College London, London, UK

Randall W. Myster Department of Biology, Oklahoma State University, Oklahoma City, OK, USA

Natalia Politi Insituto de Ecorregiones Andinas CONICET/Universidad Nacional de Jujuy, S.S. de Jujuy, Jujuy, Argentina

Luis Rivera Insituto de Ecorregiones Andinas CONICET/Universidad Nacional de Jujuy, S.S. de Jujuy, Jujuy, Argentina

Adriana Ruiz Postgrado en Ecología Tropical (ICAE), Facultad de Ciencias, Universidad de Los Andes, Mérida, Venezuela

Fausto O. Sarmiento University of Georgia, Athens, GA, USA

Sabrina Setaro Wake Forest University, Winston-Salem, USA

Pascual J. Soriano Laboratorio de Ecología Animal, Departamento de Biología, Facultad de Ciencias, Universidad de Los Andes, Mérida, Venezuela

Juan Pablo Suárez Universidad Técnica Particular de Loja, Loja, Ecuador

Wolfgang Wilcke Institute of Geography and Geoecology (IfGG), Karlsruhe Institute of Technology (KIT), Karlsruhe, Germany

Chapter 1
Introduction

Randall W. Myster

1.1 The Andes

The Andes or Andean Mountains ("*Cordillera de los Andes*" in Spanish) are located along the western edge of South America (Fig. 1.1). They are the longest continuous mountain range in the world—at least 7000 km—and have more volcanoes than any other mountain range. These volcanoes include Mount Chimborazo (Fig. 1.2), which has the point on Earth closest to the moon, and Mount Aconcagua, which is the highest peak in the Andes at 6962 m a.s.l. The northern end of the Andes begins in Venezuela (~10° N latitude), continues through Colombia, Ecuador, Peru, Bolivia, Argentina and ends in Chile (~57° S latitude). The longitude of the Andes is between ~70° W and ~80° W (Knapp 1991).

The Andes are not just one mountain range, but rather a succession of parallel and transverse mountain ranges—called cordilleras—along with their intervening plateaus and depressions. The distinct eastern ranges are referred to collectively as the Cordillera Oriental, and the distinct western ranges are referred to collectively as the Cordillera Occidental. The directional trend of both the Cordillera Oriental and the Cordillera Occidental are north-to-south, but the Cordillera Oriental bulges eastward to form isolated peninsula-like ranges and high inter-montane plateau regions. Researchers usually divide the Andes into the Southern Andes (the Chilean, Fuegian, and Patagonian cordilleras), the Central Andes (the Peruvian cordilleras), and the Northern Andes (the Ecuadorian, Colombian, and Venezuelan [or Caribbean] cordilleras: Oncken et al. 2006).

Although temperature generally increases northward from Tierra del Fuego in Chile (and southward from Venezuela) towards the equator, elevation, proximity to the sea, rainfall, and topographic barriers to the wind create much variety in temperature and other climatic conditions. Indeed, the Andes is a place of extremes

---

R. W. Myster (✉)
Department of Biology, Oklahoma State University,
Oklahoma City, OK, USA

© Springer Nature Switzerland AG 2021
R. W. Myster (ed.), *The Andean Cloud Forest*,
https://doi.org/10.1007/978-3-030-57344-7_1

Fig. 1.1 The Andean Mountains, located along the Western edge of South America and marked in dark gray

Fig. 1.2 "Alexander von Humboldt and Aimé Bonpland at the foot of Chimborazo" by Friedrich Georg Weitsch, 1810

where temperatures range from below freezing on the tops of mountains to very hot in the lower elevation tropical forests, and precipitation ranges from dry in deserts to very wet in those same tropical forests. While such climatic variation obviously occurs among elevations, there are large differences among the aspects of slopes as well, for example, those that face the Pacific, those that face the Amazon basin, or those that face different directions within the Andes.

Given this large variation in environmental conditions and thus in potential "niches" (Bazzaz 1996), it is not surprising that the Andes are an area of great biodiversity. For example, about 30,000 species of vascular plants live in the Andes—roughly half of which are endemic—surpassing the diversity of any other "hotspot" on earth (Hoorn et al. 2010). Moreover, many of these species have wide distributions, for example, animals such as bear and deer which are normally found in North America exist at high elevations in the Andes, while animals such as monkeys which are normally found in the Amazon exist at low elevations (overlap: Cardelús et al. 2006). Indeed, the Andes can seem to be a microcosm of the entire Western hemisphere! Adding to this biodiversity are species introduced by human activities such as farming (e.g., horse, cow, pig, chicken). Indeed, crops commonly include potato, maize, and beans raised for local consumption, and tobacco, cotton, and coffee raised for export (see chapters in Myster 2007a). After cultivation is no longer profitable, these areas often become pastures. Finally, mining for copper, gold, and silver, as well as iron and tin, is common in the Andes of Peru, Chile, and Bolivia.

1.2   The Andean Cloud Forest

Cloud forests exist in mountainous regions all over the earth (Bruijnzeel et al. 2010). A commonly adapted definition of cloud forests are "forests that are frequently covered in cloud or mist" (Stadtmüller 1987; Hamilton et al. 1995) or "forests affected by frequent and/or persistent ground-level cloud" (Grubb 1977). The ground vegetation in these forests receive most of their moisture from fog and wind-driven precipitation (Nadkarni and Wheelwright 2000) but also get some from rainfall. These low-level clouds profoundly affect the temperature and light regime of Cloud forests as well (Mulligan and Burke 2005). Researchers have categorized cloud forests (1) at low elevation (usually starting around 1200 m a.s.l. with 25–50% moss cover on stems) where there is incipient and intermittent cloud formation, (2) at high elevation (~2500 m a.s.l with 70–80% moss cover on stems) where cloud condensation becomes most persistent, and (3) at the highest elevation (possibly ~3500 m a.s.l with more than 80% moss cover on stems) where the forest has an "elfin" physiognomy created by exposure to wind-driven fog and rain (Bruijnzeel et al. 2010). Not counting this elfin cloud forest, cloud forests are two-layered canopy systems with abundant mosses, lichens and epiphytes (Veneklaas and Van Ek 1990). Moreover, these cloud forest types blend into each other along gradients of temperature, moisture, and elevation with average rainfalls ranging from 500 to 10,000 mm/year (Hamilton et al. 1995) and an average temperature of 17.7° C (Bruijnzeel et al. 2010).

Cloud forests have the capacity to capture or strip water from clouds which may result in an increased catchment water yield compared to other vegetation types. They also have a high proportion of biomass in the form of epiphytes, fewer woody climbers than in lower altitude tropical forests, high local biodiversity in terms of shrubs, herbs, and epiphytes, high proportion of endemic species and wet, frequently water-logged soils that typically have high organic matter contents (Histosols). These volcanic soils are generally shallow with weakly developed horizons (Leptosols) or they may be drought sensitive with low water holding capacity (Regosols: Hamilton et al. 1995). Soils tend to be anaerobic, high in soil organic matter (SOM), highly acidic, and may be both nitrogen (N) and phosphorus (P) limited (Bruijnzeel et al. 2010).

Perhaps the most obvious way to investigate cloud forests is to focus on their elevational gradient (Beck et al. 2008). Since this gradient is known, direct gradient analysis can be used to explore how both abiotic (e.g., temperature, precipitation) and biotic (e.g., plant, animal, and fungi taxa) parameters vary with either increasing or decreasing altitude. For example, we know that with an increase in elevation, both net primary productivity and precipitation (crown drip) increases as does humidity, wind speed and cloudiness, while overall water use (evapotranspiration) decreases as temperature declines. In addition, litter decomposition and soil organic matter (SOM) turnover both slow down with increasing elevation but amount of SOM and the carbon nitrogen ratio (C/N) increases, limiting nutrient availability. Mineralization rates and concentrations of all macronutrients decrease with increasing elevation. With increasing elevation, height of emergent trees, tree canopy height, and number of strata decreases as does stem diameter and leaf area, while moss, epiphytes (e.g., ferns and orchids) and tree fern cover all increase. With increasing elevation, tree stems increasingly take on a crooked and gnarled physiognomy, and bamboos often replace palms as the dominant undergrowth species. Finally, biodiversity, turnover of biodiversity and endemism may reach its peak at mid-elevation due to overlap of species distributions commonly found at higher and lower elevations (Cardelús et al. 2006).

In the Andes as elsewhere, plant and animal species form statistically significant associations (Myster and Pickett 1992; Myster 2012a). While these include species associations found in deserts and grasslands (see chapters on the Andean grasslands páramo, puna, and steppe in Myster 2012b), this book is mainly concerned with forests, and forests within the Andes range from lowland forests at low elevations (bordering, and similar to, Amazonian *terra firme* forest: Myster 2016) to, premontane forests, montane forests and finally cloud forests at the higher elevations (von Humboldt and Bonpland 1807). At the highest elevations, due to low atmospheric pressure, it can take 5-min to boil a 3-min egg (author, pers. trekking experience)! Not just elevation determines the kind of forest found in the Andes, however, but also other landscape characteristics such as aspect, slope, and the nature of the substrate (Myster et al. 1997). Andean cloud forests are also defined by disturbance including natural tree-fall (Myster 2014, 2015, 2017, 2018a) and landslides (Myster and Sarmiento 1998), and human activity such as conversion to agriculture and pasture after slash and burn forest clearing techniques, and the building of roads and urban areas (Myster 2004a, 2007b, 2009, 2010a, 2012a, 2013).

The focus of this book is Andean cloud forests, which are among the most diverse ecosystems in the world (having 1/6 of all its' plant species) and are important for the hydrological and other biogeochemical cycles of large parts of the Neotropics. This includes the carbon cycle and Andean Cloud forests may show reduced deposition and other effects of global warming/climate change in the future (Deutsch et al. 2008; Chen et al. 2009). These evergreen Andean cloud forests exist in the same elevational zones as outlined above for cloud forests found elsewhere in the world, and thus can also be characterized as low elevation cloud forests, high-elevation cloud forests, and "elfin" cloud forests (Gould et al. 2006) exposed to wind-driven fog and rain at the highest elevation. In addition to biodiversity and biogeochemical cycling, Andean cloud forests are unique and important because they exist not just over a large elevational (altitudinal) gradient, but also over a large latitudinal gradient. The interaction of these two gradients (Holdrige 1967) creates incredible variation and thus allows a more complete understanding of cloud forests than was previously possible (Beck et al. 2008; Bruijnzeel et al. 2010; Hamilton et al. 1995; Nadkarni and Wheelwright 2000; Stadtmüller 1987). Chapter authors will use this framework to explore Andean cloud forest structure, function, and dynamics.

1.3 Case Study: Primary Cloud Forest at the Reserva Biologica San Francisco

My first study site was the Reserva Biologica San Francisco (RBSF: 3° 58' 30" S, 79° 4' 25" W, Bussmann 2001; Beck et al. 2008, Fig. 1.3). The reserve covers 1000 ha of the northern slopes of Cordillera de Consuelo in Zamora-Chinchipe Province, Ecuador adjacent to Podocarpus National Park. RBSF extends between 1800 m and 3150 m a.s.l and is covered by cloud forest. Soils are Dystrudepts and Haplosaprists at lower elevations, and Petraquepts and Epiaquents at higher elevations (Bussmann 2003). Temperatures range from 15° to 17° C at lower elevations to 9–11 ° C at upper elevations, and annual precipitation from 2200 mm per year at lower elevations to 5000 mm per year at upper elevations (Bussmann 2003).

In January 2019, my field assistants and I first set up and then sampled all trees at least 10 cm diameter at breast height (dbh) in ¼ ha plots at 10 different elevations (1900 m, 2000 m, 2100 m, 2200 m, 2300 m, 2400 m, 2500 m, 2600 m, 2700 m, 2800 m) across from the Rio San Francisco, next to the RBSF field station. The dbh measurement was taken at the lowest point where the stem was cylindrical, and for buttressed trees above the buttresses. Trees were identified to family, genus, and species using Martinez (2005) and the Missouri Botanical Garden website www. mobot.org as taxonomic sources. In addition, voucher samples, kept at the RBSF on-site herbarium, were compared to field samples to facilitate identification.

Fig. 1.3 A map of Ecuador showing the location of my four study sites, all in Andean cloud forest (1) Reserva Biologica San Francisco, (2) Maquipucuna Reserve, (3) Guandera Biological Station, and (4) Yanacocha Reserve. The Andean Mountains are indicated in brown and yellow

1.3.1 Curve-Fitting Floristic Patterns

Using this collected data set, the number of stems in each family was first complied at each sampling point on the elevational gradient. The familial data was then subjected to curve-fitting analysis (Wilson 1991; Guest 2013) for these four response patterns (1) a symmetric unimodal pattern, (2) a skewed unimodal pattern, (3) a linear pattern, or (4) a plateau pattern (Austin et al. 1994; Oksanen and Minchin 2002; Rydgren et al. 2003) using least-squares regression analysis after the appropriate transformation (Hill 1977; SAS 1985; Bongers et al. 1999; McCain 2005;

Marini et al. 2010), as was done to generate dominance-diversity curves (Myster 2010a). Elevation was the independent variable and number of stems was the dependent variable in all cases. In addition, two ordinations were performed on the familial data (R-type) and two ordinations on the elevational data (Q-type) (1) an R-type principal components analysis (PCA: Pielou 1984) ordination using the correlation matrix, (2) an R-type non-metric multidimensional scaling ordination (NMDS: SAS 1985), (3) an Q-type principal components analysis (PCA: Pielou 1984) ordination using the correlation matrix, and (4) an Q-type non-metric multidimensional scaling ordination (NMDS: SAS 1985; Rydgren et al. 2003). By computing both PCA and NMDS, I could compare the results to examine possible distortion of the data by PCA (a horseshoe effect: Legendre and Gallagher 2001).

There were 29 families in the data set where the most common were Melastomataceae, Lauraceae, Clusiaceae, and Rubiaceae, and the rarest were Solanaceae, Malvaceae, Annonaceae, Cyatheaceae, and Hypericaceae. In addition, the most common species were *Clusia* sp., *Nectandra membranaceae,* and *Miconia punctata.* Statistics revealed significant symmetric unimodal response curves for Aquifoliaceae, Clusiaceae, Euphorbiaceae, Primulaceae, and Podocarpaceae, skewed unimodal response curves for Lauraceae and Melastomataceae, a linear response curve for Myrtaceae, and plateau response curves for Primulaceae and Rubiaceae. The most important families defining ordination axes were Rubiaceae, Primulaceae, Euphorbiaceae, and Anacardiaceae, and within those ordination spaces, families Melastomataceae and Rubiaceae were most different. The most important elevations defining ordination axis were 1900 m, 2400 m, 2700 m, and 2800 m, and within those elevations 1900 m, 2300 m, and 2400 m were most different. The elevation pairs 2200 m/2500 m, 1900 m/2000 m, and 2600 m/2800 m were most similar, and PCA distortions were minimal. In conclusion (1) families sampled along this gradient had an individualistic pattern, (2) all four major kinds of response pattern had significant curves among these families with a symmetric unimodal pattern most common, and (3) ordinations suggested that species in the families Melastomataceae and Rubiaceae have the most dissimilar ecological roles and tolerances (niches: Bazzaz 1996) while species in most of the families may be similar.

1.3.2 Curve-Fitting Physical Structure Patterns

Using the same data set, these structural parameters were computed for each plot at each elevation: (1) the total number of tree stems, the total number of tree stems divided into four size classes: $10 \leq 19$ cm dbh, $20 \leq 29$ cm dbh, $30 \leq 39$ cm dbh and ≥ 40 cm dbh, and mean dbh, (2) tree family, genus, and species richness, (3) fishers α diversity using the formula in Fisher et al. (1943) as realized by the JavaScript program in http://groundvegetationdb-web.com/ground_veg/home/diversity_index, (4) the sum of the basal areas of all individual tree stems ($\sum \pi r^2$; where r = the dbh of the individual stem/2), (5) above-ground biomass using the formula in Nascimento and Lawrance (2002) suggested for tropical trees of these stem sizes, and (6) canopy closure using the formula in Buchholz et al. (2004) for tropical trees. These parameters were subjected to curve-fitting analysis (Wilson 1991; Guest 2013) for these four response patterns (1) a symmetric unimodal pattern, (2) a skewed unimodal pattern, (3) a linear pattern, or (4) a plateau pattern (Austin et al. 1994; Oksanen and Minchin 2002; Rydgren et al. 2003) using least-squares regression analysis after the appropriate transformation (Hill 1977; SAS 1985; Bongers et al. 1999; McCain 2005; Marini et al. 2010). Elevation was the independent variable and number of stems was the dependent variable in all cases. Curves with zero slope were not considered ecologically meaningful and not reported, even if significant.

The total tree stems increased with elevation from 772/ha (1900 m) to almost 1000/ha (2800 m), but not montonically decreasing from 2100 m to 2200 m, the smallest stem size was always most common reaching 100% of stems at 2800 m, and the next smallest size took almost all of the remaining stems. There were very few stems over 29 cm dbh, four elevations had no tree stems over 29 cm dbh and there were no tree stems larger than 39 cm dbh in any elevation above 2100 m. No elevation was family, genus or species rich and fisher's α diversity was always low. Finally, total stems and % stems $20 \leq 29$ cm dbh had significant linear curves and % stems ≥ 40 cm dbh had a significant plateau curve. In conclusion, as number of stems increased with elevation size of stems decreased, so structural parameters such as basal area and above-ground biomass (AGB) remained relatively constant. Richness and diversity were low and also relatively constant regardless of elevation.

1.4 Case Study: Primary Cloud Forest at Maquipucuna Reserve

The second study site was the Maquipucuna Reserve (MR: 0° 05' N, 78° 37' W; www.maqui.org; Sarimento 1997; Myster and Sarmiento 1998, Fig. 1.3) located 20 km from the town of Nanegalito, Ecuador. The reserve lies between 1200 m and 1800 m a.s.l. and is classified as tropical lower montane wet/cloud forest (Edmisten 1970). It has deeply dissected drainages with steep slopes and has an annual precipitation of 3198 mm (measured from Nanegal: Sarimento 1997). The temperature ranges yearly between 14° and 25 °C, with an average temperature of 18 ° C. The reserve's fertile andisol soil is developed from recent volcanic ash deposits.

1.4.1 One ha Plot: Floristics and Physical Structure Sampling

In May 2012, my field assistants and I set up a 1 ha permanent plot in a primary cloud forest at MR at an elevation of 1200 m a.s.l. We tagged and measured the diameter at breast height (dbh) of all trees at least 10 cm dbh in the 1 ha plot. The dbh measurement was taken at the nearest lower point where the stem was cylindrical

1 Introduction

and for buttressed trees it was taken above the buttresses. The tagged trees were identified to species, or to genus in a few cases, using Romoleroux et al. (1997) and Gentry (1993) as taxonomic sources. We also consulted the website of the Missouri Botanical Garden (http://www.missouribotanicalgarden.org).

From this data set floristic tables of family, genus, and species were first complied. There were a total of 18 families—Actinidaceae (24 stems), Asteraceae (1), Brunneliaceae (1), Cecropiacea (7), Fabaceae (49), Lauraceae (118), Melastomataceae (10), Mimosaceae (7), Monimiaceae (1), Moraceae (2), Myristicaceae (9), Myrtaceae (4), Piperaceae (5), Rubiaceae (11), Solaneceae (5), Tiliaceae (2), Urticaceae (12), Verbenaceae (24)—in the plot. Lauraceae was by far the most common family and all families were dominated by a small number of genera and species. The families Actinidaceae, Fabaceae, and Verbenaceae were also common, but three families (Asteraceae, Brunneliaceae, Monimiaceae) had only one stem. The families Asteraceae, Brunneliaceae, Fabaceae, Melastomataceae, Moraceae, Myrtaceae, Piperaceae, Rubiaceae, Solaneceae, Tiliaceae, Urticaceae, and Verbenaceae had a monotonic decline in stem number as stems get thicker. This was not true, however, of the families Actinidaceae, Cecropiacea, Lauraceae, Mimosaceae, Monimiaceae, and Myristicaceae. The most common species were *Erythrina megistophylla* and *Nectandra acutifolia* (Myster 2004a, 2013, 2017).

These structural parameters were generated for the plot (1) the total number of stems in the 1 ha plot, the mean and maximum among those stems, and the total number of stems divided into four size classes: $10 \leq 19$ dbh, $20 \leq 29$ dbh, $30 \leq 39$ dbh and ≥ 40 dbh, (2) the stem dispersion pattern (random, uniform, clumped) computed by comparing plot data to Poisson and negative binomial distributions using Chi-square analysis and, if clumped, greens index was also computed to access degree of clumping (Ludwig and Reynolds 1988), (3) canopy closure using the formula in Buchholz et al. (2004) for tropical trees with the resulting percentage of the 1 ha plot area closed, (4) total basal area as the sum of the basal areas of all individual stems (Πr^2; where r = the dbh of the individual stem/2), and (5) above-ground biomass (AGB) using the formula in Nascimento and Lawrance (2002) suggested for tropical trees of these stem sizes. There were 294 total tree stems in the plot, with 153 stems between 10 and 19 cm dbh, 82 stems between 20 and 29 cm dbh, 46 stems between 30 and 39 cm dbh, and 13 stems 40 cm dbh or greater. The mean stem size was 23.2 cm dbh. Family richness was 18, genus richness 24, and species richness 25. Basal area was 11.2 m^2, above-ground biomass (AGB) 198.4 Mg, and canopy closure was 46.3%.

1.4.2 One ha Plot: Plant-Plant Replacements over Six Years

The 1 ha plot was resampled in 2014 and in 2016, using the same protocol. The data was examined for plant–plant (here tree–tree) replacements happening over time, which define plant community dynamics (Myster 2010b, 2012c, 2018b). Each living tree had a location within the plot, a species, a dbh, and a neighborhood space defined as a circular area with the tree at the center and the radius = 25 × dbh/2 (Canham et al. 2004), making the minimum neighborhood space for each sampling a circle of diameter 1.25 m. This definition of neighborhood space captured the effect of shading/canopy cover of each tree. The neighborhood space is graphed for each living tree in the 2012 sampling, in the 2014 sampling, and in the 2016 sampling (Fig. 1.4).

Comparing the 2014 sampling to the 2012 sampling, (1) 12 trees died of which 10 were in the first size class and 2 were in the second size class, (2) there were 14 new trees all in the first size class, and (3) for the rest of the trees 20 went from size class 1 to size class 2, 5 went from size class 2 to size class 3, and 2 went from size class 3 to size class 4. There were no trends due to species identity or spatial location. Changes in the neighborhood spaces suggest that there were 3 none → one replacements (Table 1.1), 4 one → none replacements, 3 one → one replacements, 1 one → many replacement, and 1 many → one replacement. Theory suggests that there were probably many none → none replacements (Myster 2018b).

Comparing the 2016 sampling to the 2014 sampling, (1) 13 trees died of which 9 were in the first size class, 3 were in the second size class and 1 was in the third size class, (2) there were 13 new trees all in the first size class, and (3) for the rest of the trees 22 went from size class 1 to size class 2, 7 went from size class 2 to size class 3, and 2 went from size class 3 to size class 4. There were no tends due to species identity or spatial location. Changes in the neighborhood spaces suggest that there were 4 none → one replacements (Table 1.1), 1 none → many replacement, 4 one → none replacements, 2 one → one replacements, and 2 many → one replacements. Theory suggests that there were probably many none → none replacements (Myster 2018b).

In the replacements found in both temporal sequences, size of stem seemed to be a more important factor than species. Species replacement patterns followed the ranking of species due to abundance—that is, the more common a species is the more likely it would be in replacements. This is supportive of the neutral theory (Hubbell 2001) of plant community dynamics that would preserve species ranking of abundance, and thus species richness over time. Such rankings may be, however, the result of niche-specific recruitment mechanisms (Myster 2018b). More sampling is needed to explore these patterns and so I have replicated this plot and sampling protocol in igapó forest (Myster 2019).

1.4.3 Closed-Canopy Cloud Forest vs. Tree-Fall Gaps: Recruitment Experiments

Tree seeds put out in closed-canopy cloud forest and in tree-fall gaps showed that seed predation took more seeds than seed pathogens or that germinated for all tree seed species and in both closed-canopy forest and in tree-fall gaps. In addition, seed predation was greater in the closed-canopy forest than in the tree-fall gaps, but this

1 Introduction

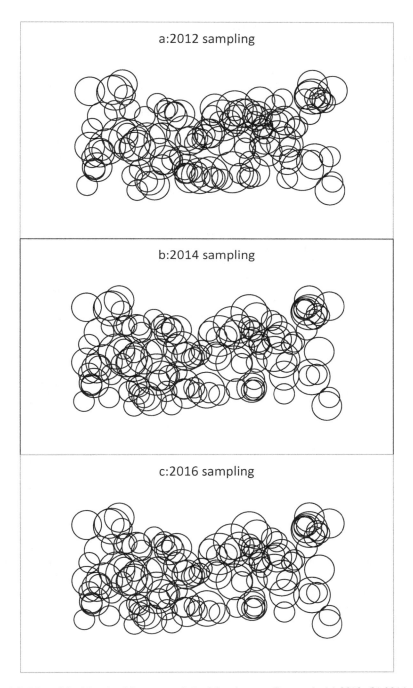

Fig. 1.4 Map of the 1 ha cloud forest sampled at Maquipucuna Reserve in (**a**) 2012, (**b**) 2014, and (**c**) 2016. Circles indicate the location and size of the neighborhood spaces of all living trees. For some trees, their neighborhood spaces extended outside of the plot

Table 1.1 The nine classes of plant–plant replacement (from Myster 2018b)

Replacement class	Description and example
None → none	No plant died and no new plant joined the neighbourhood. New phyto-space and/or new neighborhood space may, however, have been released by a pre-existing plant and reoccupied by the same or another pre-existing plant. For example, when a branch falls off a tree and another tree grows a branch into that space(s)
None → one	No plant died and one new plant joined the neighborhood. The new plant may reoccupy space(s) that was released by a pre-existing plant. For example, when a branch falls off a tree and a tree seedling recruits into that space(s)
None → many	No plant died and more than one new plant joined the neighborhood. These new plants may reoccupy space(s) that was released by a pre-existing plant. For example, when a branch falls off a tree and more than one tree seedling recruits into that space(s)
One → none	One plant died and no new plant joined the neighborhood. The space(s) released by the now dead plant may be reoccupied by a pre-existing plant. For example, when a tree dies and another tree grows a branch into that space(s)
One → one	One plant died and one new plant joined the neighborhood. The space(s) released by the now dead plant may be reoccupied by the new plant. For example, when a tree dies and another tree recruits into that space(s)
One → many	One plant died and more than one plant joined the neighborhood. The space(s) released by the now dead plant may be reoccupied by the new plants. For example, when a tree dies and more than one tree recruits into that space(s)
Many → none	More than one plant died and no new plant joined the neighborhood. The combined or collective space released by the now dead plants may be reoccupied by a pre-existing plant. For example, when a tree dies and as it falls it pulls down another tree with it. Then another tree grows a branch into that collective space
Many → one	More than one plant died and one new plant joined the neighborhood. The combined or collective space released by the now dead plants may be reoccupied by this new plant. For example, when a tree dies and as it falls it pulls down another tree with it. Then another tree recruits into that collective space
Many → many	Can be decomposed into the other replacement classes

trend was reversed for seed pathogens. Most seeds that survived, germinated. *Cecropia* sp. seeds in the tree-fall gaps and *Otoba gordoniifolia* seeds in both closed-canopy and tree-fall gap were most significantly different among all treatments (Myster 2015, 2018a).

1.5 Case Study: Secondary Cloud Forest at Maquipucuna Reserve

Within MR are secondary cloud forests of recovering sugarcane plantations, banana plantations, and pastures seeded with the grass *Setaria sphacelata* (Myster 2014; Sarimento 1997). In June of 1996, I started a long-term study by selecting (as suggested by local field assistants) six just abandoned agricultural fields: two recent sugarcane (*Saccharum officinarum*) plantations, two recent banana (*Musa* sp.) plantations, and two recent pastures seeded in *Setaria sphacelata* all at an elevation of 1300 m a.s.l. In each field, 25 5 m × 2 m contiguous plots were laid out making all six fields 250 m² rectangles. They were located within a few hundred meters of each other and had the 25 m plot border next to primary cloud forest. The plots did not have any remnant trees or sprouting tree roots at the beginning of the study, and their tree seed bank was very small. Starting in 1997 each of the 5 m × 2 m subplots of each of the six plots were sampled annually to identify each plant species and sample its percent cover—an indication of a species' ability to capture light and, therefore, to dominate these areas in the process of becoming forested communities—estimated visually in relation to each plot's area. We also sampled the diameter at breast height (dbh) of each tree stem at least 1 cm. Maquipucuna plant taxonomists, trained at the University of Georgia, USA, where voucher specimens are kept on file, assisted in the identification of species by using specimens located on site. They also used Romoleroux et al. (1997) and Gentry (1993) as taxonomic sources, as well as consulting the website of the Missouri Botanical Garden (http://www.missouribotanicalgarden.org).

These plots are part of the longest and largest old field study in the Neotropics (Myster 2004a, b, 2007a, b, 2009, 2010a, b, 2012a, 2014, 2015, 2017, 2018a) funded by the US National Science Foundation as part of the long-term ecological research (LTER) program at the University of Puerto Rico (see Myster 2013 for details). All the data is archived as LTERDATB #97, LTERDATB #100, LTERDATB#101, and LTERDBAS#109 at the LTER website (http://luq.lternet.edu). The LTER program has been a continuous part of the University of Puerto Rico and the U.S. Forest Service in Puerto Rico since 1988.

1.5.1 Permanent Plot Sampling After Sugarcane Cultivation, Banana Cultivation and Use as Pasture

Permanent plots in sugarcane fields showed grass dominating early (mostly sugarcane itself, *Saccharum officinarum*) and then declining to 50% cover levels, with *Panicum* spp. and *Brachiaria subquadripara* also present, ferns (e.g., *Nephrolepis* sp., *Thelypteris deltoidea*) and forbs (e.g., *Desmodium* sp., *Bidens pilosa, Lantana camara*) as a small part of the flora that begins to decline immediately with cover of woody species increasing steadily, trees showing this gradual increase with members of the family Melastomataceae most common, and *Piper aduncum* more common here than in other fields (Myster 2007b). In sugarcane fields, *Acalypha platyphylla*, *Piper aduncum*, and *Vernonia pallens* were common. Further, *P. aduncum* was more common in sugarcane fields than in other fields, and there were 18 tree species in sugarcane fields and 21 in banana fields. For the sugarcane fields, total cover, an indicator of stratification and developing structure, was greater compared with the other fields, and banana greater than pasture, starting at 200% cover and gradually declining with time. Species richness was also greater in the sugarcane fields compared with the other fields, and banana had more species than did pasture.

In the plots located in recovering banana fields, grass and *Musa* sp. dominate early and then decline, but to greater cover levels than in the sugarcane. Ferns attain greater cover levels in banana than in sugarcane but forbs are a small part of the flora. Also, woody species increase, but at smaller cover levels than in sugarcane. There were many of the same tree species in banana fields as in sugarcane fields, but with additional species (e.g., *Cecropia monostachya*) present. *Piper aduncum* was frequent in the permanent plots. Total stems were slightly more in the banana fields compared with the sugarcane fields, and pastures had few stems. Compared with banana, mean stem height was also greater in the sugarcane and increased through time, while pasture stems remained small. Basal area stabilized at 500 cm^2 for sugarcane and banana fields.

Finally, in the seeded pastures, grass dominated almost completely (mainly the planted grass *Setaria sphacelata*), with a few woody plants showing in year five. Fern and forb cover was small, and pastures had few trees, but the species they did have were common to the sugarcane and banana fields (Myster 2007b). There were life-form changes such as a domination by grass (e.g., *Panicum* sp., *Axonopus compressus*) that peaks in cover in year three and then declines, fern (*Nephrolepis* sp., *Thelypteris deltoidea*) and herbaceous (e.g., *Commelina* sp., *Desmodium adscendens*, *Clidemia hirta*) cover that also peaks in year three but at lower cover levels than grass, woody vines (e.g., *Ipomea* sp.) that continue to increase in cover over the first 5 years, and trees and shrubs that enter, peak, and decline in an individualistic pattern. Community parameters in pasture plots show after 5 years (1) two or three strata or horizontal levels that developed quickly, (2) total stems at 200/250 m^2 plot, (3) tree richness constant at 20 species, (4) tree-stem evenness approximately 0.7, (5) productivity peaking at year two at 0.5 kg/m^2/year and then declining slightly, (6) turnover decreasing over the first 5 years to 50%, and (7) mean height at 250 cm and total basal area of stems at 1000 cm^2.

Data in each of the 25 subplots were then pooled to tabulate a total percent cover for each species for the entire plantation or pasture per sampling year. Those total percent cover data were first sorted in decreasing order, then log10-transformed, and finally plotted for the first 20, most common, species to create dominance-diversity curves for the sugarcane and banana plantations and pastures, for each sampling year. Sampling data were then fitted to Preston's log-normal model, MacArthur's broken stick model, and both the geometric and the harmonic series using least-squares regression after the appropriate transformation. I found that the sugarcane and banana plantations had a significant log-normal pattern for the first

5 years and then flattened out. The plots in pasture, however, showed a significant geometric pattern over the first 7 years of succession. Pastures lag behind plantations and do not resemble them within this sampling time frame (Myster 2010a).

Total cover, an indicator of stratification and developing structure, was greater in the pastures than in the other fields. The cover started at 200% in pastures and gradually declined. Species richness was greater in sugarcane fields than in the other fields, with more species in pastures than in banana fields. The total number of stems was greatest in sugarcane fields and pastures, whereas mean stem height was greatest in the pastures. Basal area increased to 500 cm^2 in sugarcane and banana fields but remained low in pastures.

Further analysis of the plot data showed that the plots in Ecuador plantations show a significant log-normal pattern for the first 5 years with a flattening out after that. The plots in Ecuador pasture, however, show a significant geometric pattern over the first 7 years of succession. Pastures lag behind plantations and do not resemble them within this sampling time frame. I also found that species richness had a significant positive relationship with productivity, where the slopes of the regression lines decreased over time for all fields taken together suggested the early upswing, leveling off, and later downswing of a unimodal curve; and this unimodal pattern held true both in the recovering sugarcane (*Saccharum officinarum* L.) and in the recovering Banana (*Musa* sp.) plantations. There was a delay, however, in the development of the unimodal pattern in the seeded pasture, perhaps due to root competition between the residual tussock grass (*Setaria sphacelata*) and Neotropical tree seedlings (Myster 2009).

Finally I used the plot data to investigate significant positive associations between pairs of old field species were first computed and then clustered together into larger and larger species groups. I found that no pasture or plantation had more than 5% of the possible significant positive associations, clustering metrics showed groups of species participating in similar clusters among the five pasture/plantations over a gradient of decreasing association strength, and there was evidence for repeatable communities—especially after Banana cultivation—suggesting that past crops not only persist after abandonment but also form significant associations with invading plants (Myster 2013).

1.5.2 Recruitment Experiments in Recovering Sugarcane Fields, Banana Fields and Pastures

Seed rain sampling showed sugarcane fields had twice the number of dispersed seeds as banana fields, and 20 times the number in pasture. Seedlings germinating from soil samples were low in all fields and pastures, with a few species found in multiple fields/pastures. Seeds put out in these same six fields showed that most seeds were lost to predators, most of the remaining seeds were destroyed by seed pathogens, and most of the seeds left germinated (Myster 2004b). There was more predation in the forest compared to the old field/forest border and more seed disease away from the forest compared to in it. Tree seeds put out in sugarcane fields, banana fields and pastures showed that seed predation took more seeds that seed pathogens or that germinated in all fields and pastures, but this was less that in primary closed-canopy forest and tree-fall gaps (see Sect. 1.4.3). Pathogens took most of the remaining seeds and most seeds that survived, germinated. *Solanum ovalifolia* seeds (only species that had seeds germinate without any losses to pathogens) in banana fields and *Piper aduncum* in all fields and pastures were the most significantly different among all treatments. Insect predation was lower in the forest border microsite for *P. aduncum*, seed disease greater at 10 m from border. All planted seedlings died in the pasture, 25% survived in banana and 15% survived in sugarcane, and there was less growth away from the forest border (Myster 2004b). Results taken together with primary forest suggest that forests may recover faster after human disturbance (here agriculture) than after natural disturbance (here tree-fall: Myster 2015).

1.5.3 Permanent Plot Sampling and Recruitment Experiments in a Recovering Landslide

Plant cover in a permanent landslide plot showed four fern families and 20 vascular plant families, with species in Asteraceae, Melastomataceae, and Poaceae most common. PCA showed that plots separated best on axes defined by the families Cecropiaceae, Urticaceae, Melastomataceae, Papilionaceae, Asteraceae, and Araceae and clumping of families in PCA space suggesting common successional strategies. In two landslides, 1304 seeds were trapped from the seed rain, mainly in the family Asteraceae. Germinated soil samples from the same landslides had 475 seedlings including nonvascular and vascular families where species in Asteraceae and Piperaceae dominated. Further, ordination showed that (1) for seed rain, seed-propagule pool, and plant cover, spatial variation was dominated by differences between the two landslides rather than within-landslide plot differences and (2) the combined seed rain and seed pool data could predict the percent cover of the family Verbenaceae and that the current plant cover families could predict Asteraceae seeds and seedlings (Myster and Sarmiento 1998).

1.6 Case Study: Primary Cloud Forest at Guandera Biological Station

The third study site was the Guandera biological station (GBS: 0° 36' N, 77° 42' W: www.jatunsacha.org/guandera-reserve-andbiological-station: Bader et al. 2007; Nierop et al. 2007, Fig. 1.3) situated on the inner flank of the eastern Cordillera at approximately 11 km from the town of San Gabriel in northern Ecuador. The area is of volcanic origin and has deep dark humic Andosols developed in old volcanic ashes. Annual precipitation is 1700 mm. Diurnal temperature fluctuations range from 4 to 15 °C but annual temperature fluctuations are low (monthly means of maximum temperature vary between 12 and 15 °C: Bader et al. 2007). Within the GBS, primary cloud forest occurs between the agricultural areas, mainly potato (*Solanum tuberosum* L.) cultivation below 3300 m a.s.l., and the páramo grasslands above 3640 m a.s.l. The GBS consists of upper montane cloud forest at lower elevations and sub-alpine dwarf forest at higher elevations. The upper montane cloud forest is dominated by the trees *Clusia flaviflora* Engl. (Clusiaceae), *Ilex colombiana* Cuatrec. (Aquifoliaceae), *Weinmannia cochensis* Hieron. (Cunoniaceae), *Miconia tinifolia* Naudin (Melastomataceae) and *Gaiadendron punctatum* (Ruiz and Pavón) G.Don. (Loranthaceae) and the epiphytes *Tillandsia* sp. (Bromeliaceae) and *Blechnum schomburgkii* (Klotzsch) C.Chr. (Blechnaceae). The sub-alpine dwarf forest is dominated by the trees *Miconia tinifolia* and *Weinmannia cochensis*, a shrub *Gaiadendron punctatum*, a fern *Blechnum schomburgkii*, and a bamboo *Neurolepis aristata* (Munro) Hitchc. (Poaceae).

1.6.1 Closed-Canopy Forest vs. Tree-Fall Gaps: Recruitment Experiments

Ten primary cloud forest areas were selected randomly at GBS in May 2015. Five had new tree-fall gaps between the normal range of 100–300 m^2 in area (Brokaw 1982) and five were in closed-canopy forest. Ripe fruits were collected by hand (using gloves to reduce transfer of infections or human odor) from one local individual of *Solanum stenophyllum* Bitter (Solanaceae: bird dispersed), one local individual of *Clusia flaviflora* Engl. (Clusiaceae: mammal dispersed), and one local individual of *Palicourea amethystina* (Ruiz & Pav.) DC. (Rubiaceae: bird dispersed). Seed species are ordered by increasing seed mass (www.data.kew.org/sid/). Seeds were examined for damage using a dissecting microscope and by floating them in water. Visually damaged seeds and those that floated were discarded. Ten viable seeds of each of the three test species were then placed in three separate plastic 9-cm-diameter Petri dishes spaced 50 cm apart and placed in the center of each closed-canopy forest area and in the center of each tree-fall gap. Three dishes per gap × ten gives a total of 30 dishes. After 2 weeks in the field, losses due to seed predation or pathogens were counted and the remaining seeds were tested for germination.

In both closed-canopy forest and tree-fall gaps, *S. stenophyllum* Dunal seeds suffered the greatest losses to predators, *P. amethystina* seeds had the greatest germination and *C. flaviflora* seeds had the greatest losses to pathogens. Comparison with data from MR primary cloud forest (Sect. 1.4.3) and secondary cloud forest (Sect. 1.5.2) showed the following: (1) *Solanum* sp. suffered the greatest losses for seeds lost to seed predators in general but *Cecropia* sp. and *Ficus* sp. also had high losses at MR primary cloud forest; (2) for seeds lost to pathogens, species that lost the most seeds were unique to each study site: *Clusia flaviflora* seeds at Guandera primary cloud forest, *Cecropia* sp. seeds at MR primary cloud forest, and *Piper aduncum* L. seeds at MR secondary cloud forest; (3) for seeds that germinated the most, species were again unique to the study site: *Palicourea amethystina* seeds at Guandera primary cloud forest, *Otoba gordoniifolia* (A. DC.) A.H. Gentry seeds at MR primary cloud forest, and *Solanum ovalifolium* Dunal seeds at MR secondary cloud forest; and (4) in general, forest types differed significantly for both seed predation and seed pathogens.

Within the primary cloud forest at MR, there was a significant difference among tree seed species for pathogens and a significant difference among the tree seed species for germination, and within the secondary cloud forest at MR, there was a significant difference among tree seed species for pathogens. As elevation increases in primary cloud forests, the proportion of seed that germinates remain largely constant, but the major seed loss shifts from being due to predators to being due to pathogens. Conversion to agriculture also leads to seeds mainly lost to predators, but individual species loss levels depended on what crop had been planted previously (Myster 2018a).

1.7 Case Study: Primary Cloud Forest at Yanacocha Reserve

The last study site was Yanacocha Reserve (YR: 0° 07' S, 78° 35' W: http://fjocotoco.org/reserves-yanacocha, Fig. 1.3) managed by Fundacion Jocotoco and supported by a World Land Trust land purchase and Carbon Balanced funding is located on the northeastern slope of the Pichincha Volcano about 45 min northwest of Quito along the old Nono-Mindo Road on route to the Mindo Valley. The reserve was established in 2001 to protect the black-breasted Puffleg (Eriocnemis nigrivestis) whose known range is restricted to the Pichincha Volcano. The YR is mainly high-elevation elfin *Polylepis* sp. forest.

1.7.1 One ha Plot: Floristics and Physical Structure Sampling

In May 2015, one 2500 m^2 (50 m × 50 m) plot was established in YR primary cloud forest at 3400 m a.s.l., in a random location suggested by local field assistants, and measured using the exact protocol as in previous study sites. Voucher specimens, kept on file at the University of Georgia USA, assisted plant taxonomists in species identification. They also used Romoleroux et al. (1997) and Gentry (1993) and consulted the web site of the Missouri Botanical Garden (www.mobot.org). Data collected from all plots were used to compile floristic tables of family, genus, and species. Also calculated were (1) the total number of stems, the mean dbh among those stems, and the total number of stems divided into four size classes: $10 \leq 19$ cm dbh, $20 \leq 29$ cm dbh, $30 \leq 39$ cm dbh, and ≥ 40 cm dbh; (2) total basal area as the sum of the basal areas of all individual stems ($\pi * r^2$; where r = the dbh of the individual stem/2); (3) above-ground biomass (AGB) using the formula in Nascimento and Lawrance (2002) and suggested for tropical trees of these stem sizes; and (4) canopy closure using the formula in Buchholz et al. (2004) for tropical trees.

Palicourea sp. was the only species found at YR primary cloud forest and at MR primary cloud forests (see Sect. 1.4.1); *Vernonia pallens* Sch.Bip., *Erythrina megistophylla* Diels, *Nectandra* sp., and *Miconia* sp. were found in both primary and secondary plots at MR (see Sect. 1.5.1) and *Miconia* sp. was the only species in common between the MR secondary plots and the primary plot at YR. The mean stem size was similar between the primary MR plots and the YR plot, but the YR plot had more total stems and more stems in each size category, which lead to more basal area, above-ground biomass, and canopy closure at YR compared to MR. In the secondary plots, there were no stems larger than 29 cm dbh at breast height which lead to a much smaller mean stem size and lower basal area, above-ground biomass, and canopy closure compared to the primary plots at both sites. For the primary cloud forest at MR, an increase in elevation changed the species-level floristics more than conversion to and then abandonment from agriculture; however, while a rise in elevation increased the number of stems, agriculture reduced stem size structure. The plot at YR had a stem density of 193 stems/ha, with 91 stems between $10 \leq 19$ cm dbh, 56 stems between $20 \leq 29$ cm dbh, 24 stems between $30 \leq 39$ cm dbh, and 22 stems >40 cm dbh, Mean dbh was 23.1 cm, family richness was 11, genus richness was 13, species richness was 13, basal area was 5.2 m^2, above-ground biomass was 87.2 Mg, and canopy closure was 65.2% (Myster 2017).

1.8 Conclusions

In primary Andean cloud forest, as elevation increased the plant families Actinidacaea and Fabaceae were gained and the plant family Clusiaceae was lost. This was also seen when latitude increased. In primary Andean cloud forest, as elevation increased the number of tree stems increased while the size of the stems decreased. This lead to other structural parameters being fairly constant. Richness and diversity were low everywhere. This was also seen when latitude increased. Andean cloud forest recruitment was dominated by seed predation and its sources of variation. Seed pathogens could play a secondary part in some cases.

Secondary Andean cloud forest starts different from primary Andean cloud forest in both floristics and physical structure, but with time begins to resemble it, first floristically then structurally. This process is slower after landslides, compared to sugarcane and banana fields, pasture is between these two cases. Recruitment mechanisms are very similar to primary cloud forest in relative importance.

About this Book

Because cloud forests in the Andes are among the most important ecosystems of the Neotropics, they demand further study in order to better prepare for our shared human future. Here I take advantage of my 25+ years in the Andean cloud forests of Ecuador working with professors, researchers, local students, and technicians at four difference study sites (Maquipucuna Reserve, Guandera Biological Station, Yanacocha Reserve, Reserva Biologica San Francisco), with assistance from the LTER program at the University of Puerto Rico (http://luq.lter.network), to edit this book. One may check my personal website (www.researchgate.net/profile/Randall_Myster) for a complete listing and hard copies of those Andean cloud forest publications. Although this will be the first research book focused exclusively on cloud forests in the Andean mountains—and thus its chapter scope and detail is unique—there have been several books written on cloud forests in the past (e.g., Stadtmüller 1987; Hamilton et al. 1995). Among these books, three recent books on cloud forests in the Neotropics (Beck et al. 2008; Bruijnzeel et al. 2010; Nadkarni and Wheelwright 2000) provide background and context for this book.

Acknowledgements I thank Arcenio Barras, Michael Dilger, Bernardo Castro, Jorge Reascos, Bert Wittenberg Rebeca Justicia, and Rodrigo Ontaneda (Maquipucuna Reserve), Jose Cando and Geovanna Coello (Guandera Biological Station), and Jessica Paccha, Edgar Dario Ramon Castillo, and Pedro Paladines (Reserva Biologica San Francisco) for their help in facilitating my research in cloud forest in Ecuador. I also thank N. V. L. Brokaw and the LTER program at the University of Puerto Rico for their support at Maquipucuna Reserve. Finally, I thank Eda Meléndez and her staff for managing the LTER data sets. Some of the research was performed under grants BSR-8811902 and DEB-9411973 from the National Science Foundation to the Institute for Tropical Ecosystem Studies, University of Puerto Rico, and to the USDA Forest Service International Institute of Tropical Forestry as part of the Long-Term Ecological Research Program in the Luquillo Experimental Forest. Additional support was provided by the Forest Service and the University of Puerto Rico.

References

Austin MP, Nicholls AO, Doherty MD, Meyers JA (1994) Determining species response functions to an environmental gradient by means of a β-function. J Veg Sci 5:215–228

Bader MY, Geloof I, Rietkerk M (2007) High solar radiation hinders tree establishment above the alpine treeline in northern Ecuador. Plant Ecol 191:33–45

Bazzaz FA (1996) Plants in changing environments: linking physiological, population and community ecology. Cambridge University Press, Cambridge

Beck E, Bendix J, Kottke I, Makeschin F, Mosandl R (2008) Gradients in a tropical mountain ecosystem of Ecuador. Springer-Verlag, Berlin

Bongers F, Poorter L, Van Rompaey RSAR, Parren MPE (1999) Distribution of twelve moist forest canopy tree species in Liberia and Cote d'Ivoire: response curves to a climatic gradient. J Veg Sci 10:371–382

Brokaw NVL (1982) The definition of treefall gap and its effect on measures of forest dynamics. Biotropica 11:158–160

Bruijnzeel LA, Scatena FN, Hamilton LS (2010) Tropical montane cloud forests: science for conservation and management. Cambridge University Press, Cambridge

1 Introduction

Buchholz T, Tennigkeit T, Weinreich A (2004) Maesopsis eminii—a challenging timber tree species in Uganda—a production model for commercial forestry and smallholders. Proceedings of the international union of forestry research organizations (IUFRO) conference on the economics and management of high productivity plantations, Lugo, Spain

Bussmann RW (2001) The montane forests of Reserve Biologica San Francisco. Erde 132:9–25

Bussmann RW (2003) The vegetation of Reserva Biológica San Francisco, Zamora-Chinchipe, Southern Ecuador—a phytosociological synthesis. Lyonia 12:71–177

Canham CD, LePage PT, Coates KD (2004) A neighborhood analysis of canopy tree competition: effects of shading versus crowding. Can J For Res 34:778–787

Cardelús CL, Colwell RK, Watkins JE (2006) Vascular epiphyte distribution patterns: explaining the mid-elevation richness peak. J Ecol 94:144–156

Chen IC, Shiu HJ, Benedick S, Holloway JD, Cheye VK, Barlow HS (2009) Elevation increases in moth assemblages over 42 years on a tropical mountain. Proc Natl Acad Sci 106:1479–1483

Deutsch CA, Tewksbury JJ, Huey RB, Sheldon KS, Ghalambor CK, Haak DC, Martin PR (2008) Impacts of climate warming on terrestrial ectotherms across latitude. Proc Natl Acad Sci 105:6668–6672

Edmisten J (1970) Some autoecological studies of Ormosia krugii. In: Odum HT, Pigeon RF (eds) A tropical rain forest. National Technical Information Service, Springfield. Chapter D-8

Fisher RA, Corbet AS, Williams CB (1943) The relation between the number of species and the number of individuals in a random sample of an animal population. J Anim Ecol 12:42–58

Gentry A (1993) A field guide to woody plants of Northwest South America (Colombia, Ecuador, Peru). Conservation International, Washington, DC

Gould WA, González G, Carrero-Rivera G (2006) Structure and composition of vegetation along an elevational gradient in Puerto Rico. J Veg Sci 17:653–664

Grubb PJ (1977) Control of forest growth and distribution on wet tropical mountains: with special reference to mineral nutrition. Annu Rev Ecol Syst 8:83–107

Guest RG (2013) Numerical methods of curve fitting. Cambridge University Press, Cambridge

Hamilton LS, Juvik JO, Scatena FN (1995) Tropical 25 montane cloud forests. Springer-Verlag, New York

Hill MO (1977) Use of simple discriminant functions to classify quantitative phytosociological data. In: Diday E, Lebart L, Pages JP, Tomassone R (eds) First Inern. Symposium on data analysis and informatics, Versailles, 7–9 Sept. 1977. Vol. 1. Institute de Recherche d'informatique et d'automatique, Domaine de Voluceau, Rocqnencourt B. P. 105, 78150 le Chesney

Holdrige LR (1967) Life zone ecology. Tropical Science Center, San Jose

Hoorn C, Wesselingh FP, ter Steege H, Bermudez MA, Mora A, Sevink J, Sanmartín I, Sanchez-Meseguer A, Anderson CL, Figueiredo JP, Jaramillo C, Riff D, Negri F, Hooghiemstra H, Lundberg J, Stadler T, Särkinen T, Antonelli A (2010) Amazonia through time: Andean uplift, climate change, landscape evolution, and biodiversity. Science 330:927–931

Hubbell SP (2001) The unified neutral theory of biodiversity and biogeography. Princeton University Press, Princeton, NJ

Knapp G (1991) Andean ecology: adaptive dynamics in Ecuador. Westview Press, Boulder

Legendre P, Gallagher ED (2001) Ecologically meaningful transformations for ordination of species data. Oecologia 129:271–280

Ludwig JA, Reynolds JF (1988) Statistical ecology. Wiley, New York

Marini L, Bona E, Kunin WE, Gaston KJ (2010) Exploring anthropogenic and natural processes shaping fern species richness along elevational gradients. J Biogeogr 38:78–88

Martinez CEC (2005) Manual de Botanica: Sistematica, Etnobotanica y Metodos de Estudio en el Ecuador. Ditorial Universitatia, Quito

McCain CM (2005) Elevational gradients in diversity of small mammals. Ecology 86:366–372

Mulligan M, Burke S (2005) DFID FRP Project ZF0216 Global cloud forests and environmental change in a hydrological context

Myster RW (2004a) Post-agricultural invasion, establishment and growth of Neotropical trees. Bot Rev 70:381–402

Myster RW (2004b) Regeneration filters in post-agricultural fields of Puerto Rico and Ecuador. Plant Ecol 172:199–209

Myster RW (2007a) Post-agricultural succession in the Neotropics. Springer-Verlag, Berlin

Myster RW (2007b) Early successional pattern and process after sugarcane, banana and pasture cultivation in Ecuador. N Z J Bot 46:101–110

Myster RW (2009) Are productivity and richness consistently related after different crops in the Neotropics? Botany 87:357–362

Myster RW (2010a) Testing dominance-diversity hypotheses using data from abandoned plantations and pastures in Puerto Rico and Ecuador. J Trop Ecol 26:247–250

Myster RW (2010b) A comparison of tree replacement models in oldfields at Hutchenson Memorial Forest. J Torrey Bot Soc 137:113–119

Myster RW (2012a) A refined methodology for defining plant communities using data after sugarcane, banana and pasture cultivation in the Neotropics. Sci World J 2012:9. https://doi.org/10.1100/2012/365409

Myster RW (2012b) Ecotones between forest and grassland. Springer-Verlag, Berlin

Myster RW (2012c) Plants replacing plants: the future of community modeling and research. Bot Rev 78:2–9

Myster RW (2013) Long-term data from fields recovering after sugarcane, banana and pasture cultivation in Ecuador. Dataset Pap Ecol 2013:468973. https://doi.org/10.7167/2013/46873. 10 pages

Myster RW (2014) Primary vs. secondary forests in the Neotropics: two case studies after agriculture. In: Forest ecosystems: biodiversity, management and conservation. Nova Publishers, New York, pp 1–42

Myster RW (2015) Seed predation, pathogens and germination in primary vs. secondary Cloud forest at Maquipucuna Reserve, Ecuador. J Trop Ecol 31:375–378

Myster RW (2016) Forest structure, function and dynamics in Western Amazonia. Wiley-Blackwell, Oxford

Myster RW (2017) Gradient (elevation) vs. disturbance (agriculture) effects on primary cloud forest in Ecuador: floristics and physical structure. N Z J For Sci 47:1–7

Myster RW (2018a) Gradient (elevation) vs. disturbance (agriculture) effects on primary cloud forest in Ecuador: seed predation, seed pathogens and germination. N Z J For Sci 48:4

Myster RW (2018b) The nine classes of plant-plant replacement. Ideas Ecol Evol 11:29–34

Myster RW (2019) Igapó (black-water) forests of the Amazon Basin. Springer-Verlag, Berlin

Myster RW, Pickett STA (1992) Dynamics of associations between plants in ten old fields through 31 years of succession. J Ecol 80:291–302

Myster RW, Sarmiento FO (1998) Seed inputs to microsite patch recovery on tropandean landslides in Ecuador. Restor Ecol 6:35–43

Myster RW, Thomlinson JR, Larsen MC (1997) Predicting landslide vegetation in patches on landscape gradients. Landsc Ecol 12:299–307

Nadkarni NM, Wheelwright NT (2000) Monteverde: ecology and conservation of a Tropical Cloud forest. Oxford University Press, New York

Nascimento HEM, Lawrance WF (2002) Total aboveground biomass in central Amazonian rainforest: a landscape-scale study. For Ecol Manag 68:311–321

Nierop K, Tonneijck GJ, Jansen FH, Verstraten JM (2007) Organic matter in volcanic ash soils under forest and paramo along an Ecuadorian altitudinal transect. Soil Sci Am J 71:1119–1127

Oksanen J, Minchin PR (2002) Continuum theory revisited: what shape are species responses along ecological gradients? Ecol Model 157:119–129

Oncken O, Chong G, Franz G, Giese P, Götze H, Ramos VA, Strecker MR, Wigger P (2006) The Andes: active subduction orogeny. Frontiers in earth sciences. Springer-Verlag, Berlin

Pielou EC (1984) The interpretation of ecological data: a primer on classification and ordination. Wiley, New York

Romoleroux K, Foster R, Valencia R, Condit R, Balslev H, Losos E (1997) Especies Lenosas (dap → 1 cm) encontradas en dos hectareas de un bosque de la Amazonia ecuatoriana. In: Valencia

R, Balslev HR (eds) Estudios Sobre Diversidad y Ecologia de Plantas. Pontificia Universidad Catolica del Ecuador, Quito, pp 189–215

Rydgren K, Okland RH, Okland T (2003) Species response curves along environmental gradients. A case study from SE Norwegian swamp forests. J Veg Sci 14:869–880

Sarimento FO (1997) Arrested succession in pastures hinders regeneration of tropandean forests and shreds mountain landscapes. Environ Conserv 24:14–23

SAS User's Guide: Statistics (1985) SAS Institute, Cary, NC

Stadtmüller T (1987) Cloud forests in the humid tropics. United Nations University, Turrialba

Veneklaas EJ, Van Ek R (1990) Rainfall interception in 2 tropical montane rain-forests, Colombia. Hydrol Process 4:311–326

Von Humboldt A, Bonpland A (1807) Essai sur la geographie des plantes – accompagne´ d'un tableau physique des re´gions e´quinoxiales, fonde´ sur des mesures exe´cute´es, depuis le dixie`me degre´ de latitude bore´ale jusqu'au dixie`me degre´ de latitude australe, pendant les anne´es 1799, 1800, 1801, 1802 et 1803. – Levrault et Schoell, Paris

Wilson JB (1991) Methods for fitting dominance/diversity curves. J Veg Sci 2:35–46

Chapter 2
Dynamics of Andean Treeline Ecotones: Between Cloud Forest and Páramo Geocritical Tropes

Fausto O. Sarmiento

2.1 Andean Critical Biogeography of Scale

In the new critical biogeography framework, Cloud Forests are conceived as scalar artifacts of both, historicity's (c.f. temporal) and spatiality's (c.f. areal) interdigitations of the socio-ecological production mountainscapes. Particularly in the tropical Andes, where the greatest concentration of Tropical Montane Cloud Forest (TMCF) ecosystems exists, most of the research endeavor has been devoted to spatio-temporal "landscape characterization" of many functional traits or ecological niches available along the verdant gradient (see Chap. 1), and to create a "sense of place" that could make them unique; they persist at present appealing to the eye and to the mind. Because of my affinity with the neotropical realm, I include most examples or case studies from the tropical and subtropical Andes, yet extreme cases of cloud forests can be pointed from the Patagonian or even the Magallanic Andes of the *Terra Australis*, where the mystical fog shrouding is as spectacular as in the Andean crescent. When possible, I use vernacular descriptors of *pueblos originarios* to describe the terms used regionally for the cloud forest ecosystem, wherever they might be located.

2.1.1 Annotations of Narrative Framework

I use the phonetic alphabet to write "*Kichwa*" words with *cursive*, and I place them within quotation marks; other foreign words, including scientific notations are also *italicized* but without marks. I use double quotation marks to denote "alternative

---

F. O. Sarmiento, PhD (✉)
University of Georgia, Athens, GA, USA
e-mail: fsarmien@uga.edu

© Springer Nature Switzerland AG 2021
R. W. Myster (ed.), *The Andean Cloud Forest*,
https://doi.org/10.1007/978-3-030-57344-7_2

meaning" or to directly cite quotes from other authors *verbatim*. Geocritical texts are written in active voice, in the first person.

This chapter intersects unorthodox angles in the geoecological study of cloud forests, one that requires readers to familiarize themselves not only with TMCF terminology and morphology of the mountain landscape (*a-la* Carl O. Sauer), but also with the alternative explanations of geocriticism of decolonial ecologies of the Andes, and the need to assume a transdisciplinary approach to understand mountain studies, namely Montology (*a-la* Jack D. Ives). With this integrative and holistic view, I attempt to guide the reader through an intellectual expedition that might conduce either to: (a) the discovery of new narratives; (b) the affirmation of biogeographical dogmas; or, (c) deter curiosity and ignite activism in favor of the preservation, conservation, restoration, and regeneration of these jungles in the clouds (or "nuboselva") of the tropandean landscapes.

At any rate, this chapter to a book on Andean Cloud Forests triggers geographic and ecological inquiry of the ecotones that is fully developed later in individual chapters. In this quest of Cloud Forests borders (*a-la* Daniel W. Gade), we shall become explorers of the mountain lines in the landscape (Fig. 2.1), where the unexpected turn, the slippery rock, the thorny pole, the mossy carpet, the hanging gardens, the misty and shrouded canopy, the waterlogged ponds, the myriad of sounds, the ubiquitous rainbows, the exquisite flavors, the poisonous stings, and the psychedelic elixirs will all conform a collection of discursive mysteries that ultimately inform and elucidate what hidden dynamics can be seen at the Andean Treeline Ecotone Region (ATER). The so-called alpine treelines (Malanson et al. 2011), following European traditions of seeing them as in the Alps (*a-la* Alexander V. Humboldt), is a line that separates forests from grasslands, a line that divides cleavages and hills, a line that draws agriculture and livestock, a line that mimics firewalls and mowers, a line, that is neither theoretical nor practical, but imagined… A line that is *al-Barzakh!* So, seat back, relax, grab your poncho, and rubber boots, clear your sunglasses, refill the insect repellent and sunscreen… and be prepared to be wowed!

2.2 Why We Call them Cloud Forests? Onomastics at Play

As the only Commandment of Geography dictates, "*Know Thy Word*" will prompt us to start the exploration of place naming with instances where scientists have used misnomers. By using etymology and onomastics, the real meaning of cloud forests words will be unveiled to better grasp the complexities of the adaptive system of the verdant (or "*Yunga*"). During the 1980s, late professor Larry Hamilton led an effort to coalesce a global network of mountain scholars interested in TMCFs, spearheaded by the institutional concerns of the East-West Center in Hawai'i, the US Forest Service and the Luquillo Forest scientists led by Fred Scatena in Puerto Rico, and James Juvik of the University of Hawai'i, Hilo, as well as catalyst and foundational research on forest hydrology by L.A. Bruijnzeel and collaborators in European and North

2 Dynamics of Andean Treeline Ecotones: Between Cloud Forest and Páramo…

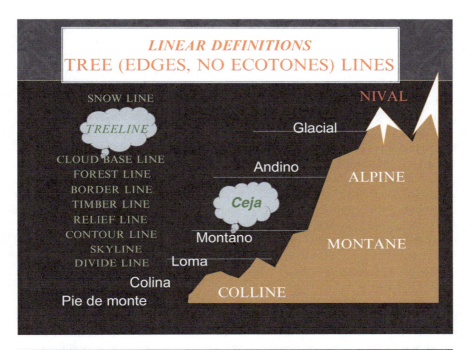

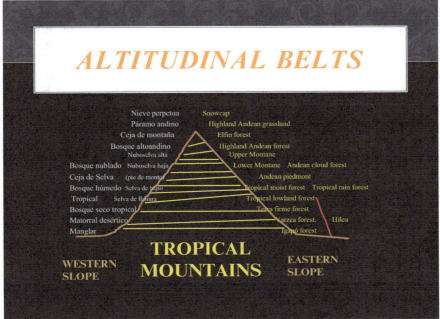

Fig. 2.1 A display of different lines found in the tropical mountainscape

American Universities (Hamilton et al. 1993). The conference proceedings, published first as a book from the East/West Center and then as special issue of *Ecological Studies*, prompted a frenzy of articles published in ensuing years. The introductory chapter of the *Ecological Studies* volume on the subject left a clear impression that it was the subject matter for botanists, forest ecologists, and hydrologists (Hamilton et al. 1994). There was a dearth of contributions from the social sciences, particularly political ecology, human geography or environmental anthropology.

That failure was partially corrected when another international conference was held at the New York Botanical Garden, mostly energized by E. Forero, S. Churchill, H. Balslev, and J. Luteyn, botanists wishing to highlight the use of the extensive yet old botanical collections (Churchill et al. 1995). This international conference in the Bronx's Gardens highlighted floristic and cladistic emphasis but it was also decorated with a few ethnobotanical works and some socioeconomic analyses of the useful cloud forest plants and the concern for the human impact on those ecosystems (Sarmiento 1995a). It was not until a global meeting was held again, this time in the big island of Hawai'i, (Bruijnzeel et al. 2011) that the human dimension and the need of management was at the core of concerned cloud forest scientists, pushing the need of integration and multidisciplinary research (Martin and Bellingham 2016).

Perhaps ignited by the imminent deforestation frontier of these orobiomes in the developing world, the plight of shrouded mountainscapes became the target of international programs, not only of biodiversity conservation and water capture, but also on other ecosystem services, hunger prevention, poverty alleviation, and the negative prospects of climate change (Foster 2001). There were programs on Cloud Forests at the IUCN, at the WCMC, at the FAO's Mountain Partnership, ICIMOD, GMBA, GLORIA, MRI, and other multinational initiatives for pushing the scientific needs of TMCFs to the next level. Furthermore, diverse NGOs were created with specific targets of working with cloud forests issues (Bubb et al. 2004). However, prolific writers on the social sciences have delved notions of manufactured landscapes, landscape archeology, hybrid socio-ecological systems, coupled nature-society relations at the forest edges with alpine treelines, and published their research in a number of professional journals. Recently, a renewed effort to bring traditional ecological knowledge (TEK) and ethnoecology makes the study of mountains and of cloud forests, a vivid front of scholarship (Taylor 2010; Sarmiento and Hitchner 2017). A special call for action is formulated (Frascaroli and Fjeldted 2018) and special panels within international conferences on Mountains (e.g. Perth III in Scotland, Mountains 2018 in Nova Friburgo, Brazil, an International conference on Past Plant Diversity, Climate Change and Mountain Conservation held in Cuenca, Ecuador 2019, and the global IMC 2019 in Innsbruck, Austria) are held.

2.2.1 Landscape Phenology and Threatening Treeline Myths

By switching the rhetoric order, we highlight these cloud forests environments because of their main characteristic: they are "forests in the clouds." Literally, they become islands in the sky, where the circadian rhythm of the incoming atmospheric tides bathing the mountain slopes at least once a day. The higher the cloud base, the more difficult to identify the treeline which is useful to separate the forest and the grassland, whether páramo proper or highland pasture. Indeed, so clear is the effect of the high tide of clouds, that several plants show the adaptations of aerial roots, even pneumatophora, just like their distant cousins in the mangrove swamps by the seashore. Horizontal precipitation affects the slope depending on the aspect of the mountain. In some cases, such as in the Atacama Desert, this phenomenon locally known as the "*kamanchaka*" provides the only source of water to support the "Lomas" vegetation and the xeric fauna therein. At present, with the advent of technological advances (e.g., netting with hydrophobic fabrics and drip irrigation), capturing the "*kamanchaka*" fog with "cloud stripping" technique has transformed those once-dry-valleys into the bread-baskets and fresh-fruit-providers of the global economies. Nowadays, avocado, cantaloupes, grapes, cranberries, and a myriad of other foodstuff have switched the economies of former abandoned mining towns in the desert, because of the trapping of the mist with physical obstacles serving as sponges to collect irrigation water, just like the leaves and branches of cloud forest trees do.

This is exemplified in the literature of the drier areas with the myth of *Garoé*, the "fountain tree" or "rain tree" (*Oreodaphne* sp.) of the Canary Islands (Gioda et al. 1995) and the Argan bush (*Argania spinosa*) with tree-climbing goats of the Moroccan mountains (Delibes et al. 2017). In the wetter confines of the Tilarán mountains of Costa Rica, the epiphytic garden (including mosses, lichens, liverworts, ferns, even tank-bromelias, and small berries) increase the capacity of the leaves and branches of the trees to act likewise in cloud stripping (Lyons 2019). It was precisely at the *Pico del Teide* in the Canary Islands that Alexander von Humboldt in 1799 first provided a mountain profile sketch of the treeline location, something that he later was able to expand to explain the different vegetation zones along the slopes of *Chimburasu* in his famous *Tableau Physique* of 1807, establishing the dogma of altitudinal zonation in mountain geoecology (Appenzeller 2019). I recall when influenced by the Humboldtian law, mountain scientists including Lauer, Acosta Solis, Budowsky, and specially Troll, developed a nuanced approach to study Andean heights under the light of mainly German scholarship that followed the altitudinal layering school (Appenzeller 2019). I also perceived the mountains segmented, determined by the climatic envelope of temperature, precipitation, and evapotranspiration. As a clear example, I include how I represented the highlands in Santa Cruz Island of the Galápagos with treelines separating the "*pampa insular*" of bracken ferns (*Pteridium aquilinum aracnoides*) of the higher reaches from the lower scrub of "*cacaotillo*" (*Miconia robinsoniana*) and la "*selva insular*" mainly of *Scalesia pedunculata* (Fig. 2.2).

These descriptive views were soon challenged by a cadre of biogeographers included Hans Ellenberg (1958, 1979) who famously asked ecologists to inquire deeper whether forest or grasslands in the Andes exist in the upper reaches; Knapp (1991) who recalled the Andes as dynamic systems adapted through time; Baeid (1999) who brought the attention of the human driver for *Polylepis* woodlands; and Luteyn and Balslev (1992) who compiled evidence to determine the burning of páramo ecosystems as the influence of community structure and composition, and

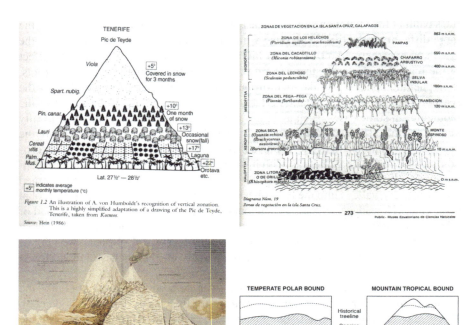

Fig. 2.2 Comparing the illustration of El Teide and Santa Cruz to illustrate the influence of the Humboldtian law in the concept of treeline

Sarmiento and Frolich (2002) who contested the naturalness of the Andean treeline, arguing for the importance of agroecology and the human dimension as drivers of the treeline dynamics in transformed farmscapes.

2.2.2 Atmospheric Tides and Fluvial Rhythms

The diurnal adiabatic warming pushes de warm water vapor towards the higher reaches, making coalescence of minute condensation nuclei, creating a mist of condensation fronts that never finish cooling enough (or "*puyu*") as to produce rain (or "*para*"). The continual push of trade winds moves ever slowly the big mass of cloudiness towards the terrain, which makes it ideal for the presence of many bryophytes that decorate every square inch of the contact surface, whether roots, stems, branches, leaves, even flowers or fruits, could harbor a massive presence of commensals amidst the small droplets of the cold, windswept mist (or "*paramuna*"). *Sphagnum*-rich ponds, bogs, fens, and rocky substrate actually act as a giant sponge

Fig. 2.3 Representation of the rain shadow effect as observed in northwestern Ecuador

sucking up more than twice their weight in water! (Nadkarni and Solano 2002). *Tillandsia*-rich branches support the increased weight of the wetted foliar mass and other epiphytic flora, contributing to the sponge-like function of the Cloud Forest "belt" (or "*yunga*") (Fig. 2.3). However, there is also the nocturnal flow of the katabatic winds that bring the load of heavy cold fog towards the valley bottom (or "*tutapuyu*"). Throughout the tropandean mountains it is very common to see the clouds accumulated down below at dawn, covering low laying cities, crops, tree plantations, and remnant forests, submerged under a "sea of clouds" (or "*puyumanta*") that disappear as soon the morning warmth returns.

2.2.3 Tide, Gravity and the Will of People

With this "intertidal" movement of fog it is obvious that the water content of both atmospheric and terrestrial interphases is saturated and remain waterlogged several days, even during dry spells of lacking rains. The arboreal life becomes stressed by the assured presence of the clouds, the limitation of nutrients and other factors associated with the fast growth, lixiviation and decay of organic matter, such that breaking of the enormous weight of the epiphytic garden is one of the main factors of ecological succession, creating canopy gaps and triggering patch dynamics and matrix regeneration on the slopes by small clearings and mini-landslides (Myster and Sarmiento 1998). This common breakage of the branches or complete falling of twisted trees, also triggers rockslides, mudslides, and other erosional processes. For instance, the "*Waiku*" of the central Andes is the phenomenon of destructive forces of lahars, rockslides, landslides, and slope failures that runs down the river bed creating havoc downstream. Excess water is not the only culprit, as sporadic or episodic it might be, but frequent tremors, even earthquakes and glacial lake out-burst flows also contribute to this geodiverse mosaic of rugged topography. However, as it has been demonstrated profusely elsewhere, the zigzagging road construction (or "*kingu*") and the switchback hiking trails (or "*kulunku*") with faulty design are among the most pernicious factors of TMCFs destruction and change (Sarmiento 1995b; Laimer 2017). Landscape architects now are reviving ancient techniques to build mountain roads without the detrimental effect of destruction, such as the "*Inka*" did to cross precipitous cliffs and forested slopes when implementing the Great Inka road (or *Kapak Ñan*) (Penney and Oschendorf 2015).

2.3 Lines in the Longer Cyclicity of Mountainscapes

As circadian rhythms mark the daily routine of back and forth cloudiness, the dif-ferential temperature is as significant as prompting extremes cold and hot measure-ments. It is common for equatorial TMCFs to exhibit the four seasons in one day. Throughout the tropics, seasonality is expressed also in the circannual rhythm but instead of having winter and summer as the opposite thermal drivers, it is the abun-dant rainfall and its complete lack thereof that pulls phenology. The Dry Season "época seca" could last several months as well as the Rainy season "época lluviosa" could be prolonged and severe. World record annual precipitation measurements have been registered near Puyo pluviometric station towards the Amazon, and also near Lita pluviometric station towards the Choco, two of the rainiest places, where people often mentioned that there are only two seasons in the year: rain and deluge (Sarmiento 1987). In these wettest places, the vigorous growth of the forest at high elevation makes it difficult to find a treeline, and one has to be inferred only from proxies or actual evidence of occurrence of "indicator" plants, as for the presence of silver-leafed *Cecropia* sp. in the lower limits, or the presence of tree fern *Cyathea* sp. in the actual belt, and the presence of *Blechnum* in the upper limits.

Conversely, in the driest extremes, cloud forest formations on the semi-desertic watersheds of the Atacama are forming the "Lomas" vegetation, with green cano-pies that form the oasis following the riverbeds. Treelines of the xerophytic pacific mountains are definitely constrained by the continuous browsing of llamas and gua-nacos, and more recently of goats and sheep mowing the slopes. On the other flank, drier areas are found at extreme northern ends whereby herbivory is also an impor-tant factor controlling the vegetation, absent of conifers but abundant of Cactacea, Leguminosae, and Burseraceae.

The curious effect of the yearly fluctuation in temperature but a rather uniform precipitation (horizontal and otherwise) has been observed as the climatic envelope that identify TMCFs with circannual rhythms; however, having virtually no change in photoperiodicity, the abundant 12 h of daily radiant energy throughout the year (either with sunlight in open blue skies, with filtered UV-ß shrouded mist or with a heightened incidence of infrared light), triggers a huge diversification in flora and fauna (Kapelle and Brown 2001) due to the proneness of mutation generated by the electromagnetic conditions in the shortest wavelength of TMCFs. This mutation-prone community shows the fastest rate of evolution due to transiliency between short-lived extinction rates, such as those observed in orchids at the Centinela farmstead in Western Ecuador (Dodson and Gentry 2001). The Centinelan extinctions have been identified as the most severe threat to biodiversity conservation in TMCFs areas, where a bulge in the distribution of life on terrestrial ecosystems makes them hot-spots for biodiversity (Myers 2003); hence the meteorological conditions of cloudiness and photoperiodicity have far-reaching biological consequences in the mountainous terrain.

Ultimately, we shall not forget the quinquennial, decadal or centennial cyclicity driven by either planetary forces, such as forming atmospheric rivers' deltas every 5 years or so; or the El Niño Southern Oscillation (ENSO) every 10 years or so; or biological forces, such as the population emergence of cycads and other insects every 20 years or so; the masting of mountain bamboo (*Chusquea* spp.) every 40 years or so; the algae bloom of cyanobacteria related to freshwater lake toxicity every century or so; or the massive hemoglobin saturation of some lakes that are dyed red (e.g., "*Yawarkucha*") every 300 years or so due to either algae bloom (*Dunaliella c.f. salina*) or the eutrophication by Halobacteria and microphytic Dinoflagellata. These larger temporal cycles create a mosaic of life in the TMCFs that is made of differential spatial components and of various time periods, where the tessellation (or how the tessellar microsites—ecotopes—are arranged) in functional traits (or cellularity of the landscape) determining either the actual or realized fabric of the Andean treeline ecotone region.

2.4 Treeline Differentiation as Response of Latitudinal Variation

It is clear that the location and extent of the Andean treeline ecotone region (ATER) changes with the latitude, due to three main factors: firstly, the incidence of direct sunlight in northern or southern exposures after the tropics of Cancer or Capricornio will determine the presence/absence of a growing season in Spring and Summer and of a withering season for the Autumn and Winter towards the subtropical or temperate mountains. While trees from the Neartic could be distributed into these TMCFs, such as Quercus sp. into Costa Rica, the southern ecological equivalents are present from the Antarctic, such as Podocarpus sp. into Colombia. Secondly, the extent of the ATER is directly related to altitudinal verdant prompted by the location of the slope itself in relation to the mass of the mountain edifice. Throughout the southern Andes, for instance, tree line appears to be just a sharp boundary around 1800 m as many of the tallest volcanoes do not reach the 2500 m a.s.l. altitude. In fact, it is possible to walk from the glacier directly to the tall *Araucaria* forest with no apparent ecotone present at any side of the glacier mass, while in the equatorial mountains the forest composition and structure remain robust but with noticeable reduction in stature. Finally, the long extent of the Andes cordillera allows for the influence of meteorological conditions to affect the manifestation of a continuous belt of cloud forest, as it happens in the Andean crescent, or as patchily distributed among the tallest mountain peaks isolated from the longer cordillera. There is much more ruggedness towards the higher latitudes, which creates atmospheric conditions that could change violently. Microclimatic conditions generated by the presence of tall cliffs, or isolated ledgers are important consideration in the study of the distribution of TMCFs and the robustness of a distinctive ecotone in the treeline region (ATER).

The curious effect of the yearly fluctuation in temperature but a rather uniform precipitation (horizontal and otherwise) has been observed as the climatic envelope that identify TMCFs with circannual rhythms; however, having virtually no change in photoperiodicity, the abundant 12 h of daily radiant energy throughout the year (either with sunlight in open blue skies, with filtered UV-ß shrouded mist or with a heightened incidence of infrared light, triggers a huge diversification in flora and fauna (Kapelle and Brown 2001) due to the proneness of mutation generated by the electromagnetic conditions in the shortest wavelength of TMCFs. This mutation-prone community shows the fastest rate of evolution due to "transiliency" between short-lived extinction rates, such as those observed in orchids at the Centinela farmstead in Western Ecuador (Dodson and Gentry 2001). The Centinelan extinctions have been identified as the most severe threat to biodiversity conservation in TMCFs areas, where a bulge in the distribution of life on terrestrial ecosystems makes them hot-spots for biodiversity (Myers 2003); hence the meteorological conditions of cloudiness and photoperiodicity have far-reaching biological consequences in the mountainous terrain. Ultimately, we shall not forget the quinquennial, decadal or centennial cyclicity driven by either planetary forces, such as forming atmospheric rivers' deltas every 5 years or so; or the El Niño Southern Oscillation (ENSO) every 10 years or so; or biological forces, such as the population emergence of cycads and other insects every 20 years or so; the masting of mountain bamboo (*Chusquea* spp.) every 40 years or so; the algae bloom of cyanobacteria related to freshwater lake toxicity every century or so; or the massive hemoglobin saturation of some lakes that are dyed red (e.g., "Yawarkucha") every 300 years or so due to either algae bloom (Dunaliella c.f. salina) or the eutrophication by Halobacteria and microphytic Dinoflagellata. These larger temporal cycles create a mosaic of life in the TMCFs that is made of differential spatial components and of various time periods, where the tessellation (or how the tessellar microsites—ecotopes—are arranged) in functional traits (or cellularity of the landscape) determining either the actual or realized fabric of the treeline ecotone region.

2.5 Mountain Effects and the Andean Treeline Ecotone Region

Whether in América, Africa or Asia, the timing of the intertropical convection zone (ITCZ) drives the permanence of the cloud forest belt, yet the boundary of forest/pasture received the name from Europe: Alpine treeline. In this chapter, I will follow Acosta-Solís's advice to emphasize the use of "Andean" treeline instead of "Alpine" treeline for the importance of highlighting unorthodox models followed in the Neotropics that contest the Humboldtian paradigm of altitudinal belts (Acosta-Solis 1976; Sarmiento 2000, 2002; Varela 2008). This is reflected in the orientation of the forested slopes, as the rainshadow effect is critical in the presence/absence of the biota exposed to horizontal precipitation along the longitudinal extend of the Andes. An obvious case is that people live on high mountains in the tropical Andes and they do not in the Alps. Similarly, depending on where in the tropics the TMCF exists, you will find significant differences in the location of the cloud base, and thus, the treeline ecotone in the Central American or Caribbean páramos, the Mexican or Guatemalan zacatonales, the Venezuelan, Colombian and northern Ecuadorian páramos, the southern Ecuadorian and northern Peruvian jalca, the southern Peruvian, Bolivian, northern Chilean and Argentinian punas, and southern Chilean and Argentinian meadows.

Slope and aspect determine the location for Alpine treeline in north/south differential exposure, while in the Andean treeline is west/east exposure that determines the ATER location. For instance, when exposed to the rugged topography, plants of the exposed slopes can experience strong lateral winds, in the Venturi effect, that prompt canyon lands and cliff faces, including ledges, to have flagged trees with obvious morphology impacted by lateral winds along the brooks, whether coves "rinconadas" or meadow-like "esteros"; however, if the strong winds are adiabatic and move vertically from the bottom of the hill towards the summit, the Bernoulli effect takes places, pushing the canopies closer to the ground, and the wind speed exerts this continual forcing towards the summits to "comb" the leaves and twist the branches, typical of the elfin forests. Depending on the force of the wind due to its velocity, some extreme cases are observed, generating the *Krumholtz* of short stature and ground hugging stems or "*chaupicaspi*" typical of the mountain bamboo (*Chusquea* sp.), the mountain palms (*Geonoma* sp.), and the broadleaf *Gumnera* sp. Further up, plants adopt the strategy of clumping to grow as rosette (e.g., "Frailejón," "almohadilla," or "romerillo"). In both cases, due to intense wind flow, desiccation of the foliar tissue is prevented by developing coriaceous leaves and spines, so that the system is faked to function as in desert physiology, despite being waterlogged.

Known also as the telescopic effect, the "massenerhebung" effect correlates the size of an island and the distance from continental areas to the lower/higher altitudinal belt of TMCFs. It is not surprising that we find insular jungles of TMCFs on isolated islands in the Galápagos archipelago at 500 m a.s.l., and the equivalent formation at 2500 m a.s.l. in the continental cordillera more than 600 nautical miles away. Whether be in the Western flank (c.f., Cisandean domain) or in the Eastern flank (c.f., Transandean domain) both effects determine the harsh xerophytic semi-desertic slope on the leeward, while the abundant forest cover brings the hydrophytic hilea on the windward. Countless records of this differential humidity due to the Phön effect in the "rainshadow" and the mass/energy effect created the difference between the enigmatic layering of mountains (Rahbek et al. 2019) and the altitudinal differentiation of the ATER on outer slopes of tropandean landscapes, hanging valleys, and "ceja de selva" (or "*yungas*") being saturated of water and plant life, while the inner slopes toward the drier Interandean valleys, harbor "bolsones," "hoyas," "vegas," and "pampas" (c.f., Interandean domain) where little "natural" vegetation could be used to draw a treeline.

Other effects are also found in TCMFs, including the little studied Cascading Effect. I am not referring to the actual physical phenomenon of downslope movement, sometimes catastrophic, related to rocksiles, mudslides, landslides or other forms of mass movement. The trophic cascades generated by introduction of some animal species have to be better understood. For instance, prior to the declaration of the Chimborazo Faunal Reserve in 1976, no vicuña existed in the area. After a successful pilot project and then a concerted effort to promote valuable wool trade with the communities nearby, the image of the "natural" paramo with the iconic camelids is now used to promote ecotourism. An important cascade effect detected with the introduction of the black fly and the ubiquitous effect of voracious grazers such as sheep, swain, cattle. Free roaming cattle of the past decades have allowed the presence of feral bulls that are traditionally sough after the entertainment of the villages festivities with bullfighting.

2.6 Conclusion: Andean Treeline Ecotones: Neither Trees nor Lines

Much research has taken place to explain why the plant formations in the upper basins of mountains show the physiognomy of elfin forests, chaparral, páramos or balds (Allen et al. 2004; Young and Leon 2006; Malanson et al. 2007; Körner 2012; Mathisen et al. 2014; Wang et al. 2016; Malanson and Resler 2016; Möhl et al. 2019 and elsewhere). Most available references relate to the study of physical parameters and positivistic explanations. In the last decade, thou, a renewed effort of understanding gradients of many types, altitudinal and otherwise, from alternative points of view; social sciences, humanities, and even the arts are increasingly available in both scholarly and gray literatures (Naess 1989; Chepstow-Lusty et al. 1996; Bowman et al. 2002; Sarmiento and Frolich 2002; Varela 2008; Sarmiento 2012; Sarmiento et al. 2017; Huisman et al. 2019 and elsewhere).

There are two emergent themes that coincide with both fronts. Firstly, that there is no such a thing as a "line" in the treeline; the boundary mosaic tends to be an ecotone (Myster 2012) that shows different structures within, such as the disturbance edge, the saum, the drip shadow, the mantel, and the veil, that it can hardly be represented with a line. These are mountains with Andean treeline ecotone region (ATER) extending several 100 m down the of the Andean flank, exhibiting a marked ecocline in functional ecology, while others tend to be localized around cliff faces, promontories or ledges (Resler 2006; Elliott and Kipfmueller 2010; Sylvester et al. 2014). Secondly, that there is no doubt that the human impact has caused the location of the forest/pasture edge; mostly as a response to fire management and grazing intensity, the ensuing boundary mosaic in the ATER resembles the intended purpose of agriculture or livestock rearing instead of a natural formation following climatic envelopes (Miehe and Miehe 2000; Gehrig-Fasel et al. 2007; White 2013).

As per the structure, composition and changes of the TMCF ecosystems, composition and change, the notion of Andean treeline is often problematic, as most species in the highlands are short, woody vegetation that could fall under the typology of forest, traditionally applied to the areas where significant timber biomass and tall poles could be readily harvested. There are notable exceptions, such as the Andean pine (*Podocarpus*—cf *Retrophyllum or Decussocarpus*—*rospigliossii*) that shows its tall stature and yellowish cylindric shaft well into the upper reaches of most primary watersheds, due to its phenotypic plasticity that allows it to survive both in the lowland Amazon and the Andean highlands. However, most of ATER species exhibit asymmetrical architecture to achieve anchorage (Chlatante et al. 2003); they are not tall and their stems are not vertical, but twisted and turned, following the contour and sustained by buttresses and aerial roots. Notwithstanding physiological adaptations, most vegetation of the ATER is also affected by both topographic and meteorological stresses, showing pubescent, tomentose or spinose stems that add to the potential of capturing horizontal precipitation. Even more impressive, the display of prostrate architecture with decumbent and procumbent stems, climbers, and lianas provide a physical tension that affects overall growth and offers the short stature, twisted appearance of "krummholz" or elfin forests in the areas or chaparral, or in generally dwarf physiognomy of the ATER, as well creating a "false ground" that often makes it difficult to walk straight, as you are stepping on a fabric of superficial structures that not only capture water but also create appropriate soil biome in ephemeral knoll, hummock, dune, or tumulus on the slopes. For instance, these elfin forests and hillocks are territory for one of the largest earthworms (or "*kuika*") (*Thamnodrilus* sp.).

2.7 Discussion Remarks: Treelines of the Mindscape

The Andean treeline is a good example of the P.T. Barnum effect: most people think they know what it is, but only few can grasp the actual meaning of the word of this reified, situated, and partialized views of the ATER. The areas where Andean treeline ecotone regions are located have been described as socio-ecological, production landscapes (SEPLS). I have argued that these areas fit very well in the Satoyama Initiative, as they exemplify the hybridity of the natural and cultural factors in "manufactured landscapes" (Fig. 2.4). A short grapple with the human dimension follows to emphasize the need of Montology to really capture the essence of place of the Andean treeline (Fig. 2.5).

Environmental cognition based on Western science afford us the possibility to characterize the landscape with conventional tools of vegetation science, taxonomy, and allied disciplines. However, there is a large chapter that is present in every zone of the TMCFs that relates to the intangible properties of the mountainscape. It is not

Fig. 2.4 Representation of the socio-ecological production landscape of the *Imbabura* volcano, as exemplifying the nature/culture hybrid of the Andean treeline, hosting ancestral practices of the *Utawallu runakuna* and the modernity of mestizos of Ibarra

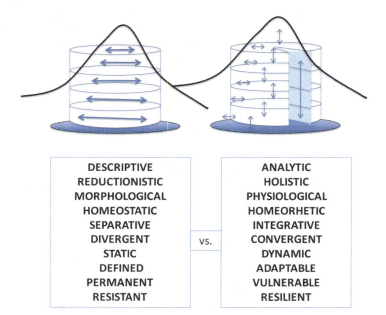

Fig. 2.5 The tendency of Montology as a transdisciplinary mountain science, shifting paradigms of the past to construct a new vision of tropandean mountains whereby the Andean treeline is contested, appropriated, and adapted to several environmental cognition pathways

only the higher elevation, the low temperature, and the excruciating humidity of the climatic envelope of cloud forests, but the appearance of the shrouded slopes, full of epiphytic growth, and udic overtones that portrait a mysterious, even dangerous perception of the ombrophilous environment of TMCFs. Phosgene, along with methane emanating from highly putrid fens due to organic decomposition in the forest floor, cause autoignition's subtle flares that prompt people to think they are seeing ghosts or mysterious lights within the canopy. Sounds coming from the undergrowth and hummock add to the magic envelope to which residents have associated metaphysical explanations.

These qualitative observations make these sites prone to superstitious and magic as the wherewithal to survival in these extreme conditions. The myths of the Cloud Forests are yet told anew, although for centuries people of the region have known about them. For instance, in areas of the *Pululahua* Geobotanical Reserve in Ecuador, near the town of Nieblí (apropos name for a town that submerges in cloudiness every day at 2:30 pm without fault) the existence of the *duende* who owns the seepage basin of *el infiernillo* that could take the unsuspecting visitor and kidnap her/him towards mysterious waterfalls and caves. There are myths that associate the presence of wild people ("*sacharuna*") that walk the boundary of the forest in the ATER, but most likely they refer to the Andean spectacled bear (*Tremarctos ornatus*) (or "*Ukumari*"), the black Andean tapir (*Tapirus pinchaque*) (or "*wagra pinchaki*"), or the Andean small deer (*Masama rufina*) (or "*sochi*") that tend to be vagrant along the crepuscular treeline.

Also, in the TMCFs of the western flank, the *aluvillo* tree (*Toxicodendron striatum*) is revered for the properties of generating rashes and violent dermatitis if the distracted pedestrian crosses under its canopy without waving the hat to salute: "*Buenos días, compadre aluvillo*"; it seems that taking the hat off the head, moves the colloidal aerosol underneath the branch of this Anacardiaceae so that the person could walk without facing this invisible barrier. In other cases, original people who inhabit these environs, seek the help of *curanderos* (or "*yachay*") so they could clean off the negative energy loaded when walking those mountain trails. All residents relate to the elements of the cloud forests with reverence, given animistic interpretation to thin air (*mal aire*), dense clouds (*espanto*), heavy downpours (or "*washishka*"), small ponds and lakes, and, of course, ice, rain, and waterfalls (Sarmiento 2016). This is still experienced by the *Kogi* people in Colombia, inhabiting the Sierra Nevada de Santa Marta National Park who are the custodians of the Lost City or "Ciudad Perdida." Just like the *Pemon* nation in *Auyantepuy* massif of Venezuela, who held sacred the tallest waterfall, *Kerepakupai Meru* (formerly Angel's fall).

Moreover, in Ecuador, the *Shwar* people define themselves as the people of the waterfall and hold therein initiation rites that include hallucinogenic chants and astral travels. Further south in Peru, along the Ceja de Selva near Chachapoyas, the *Catarata de Gocta* (the third tallest waterfall on Earth) is just one of the many flowing from the Andean heights towards the Amazon, in what is now considered the "route of the waterfalls" as a tourist destination, adding to the impressive *Kuelap* citadel (or *"llakta"*) ruins and the cliff hanging sarcophagi of the Chachapoya mummies or *pinchudos*.

Myth and superstition are still defining factors in the understanding of these little-known places. Underneath the closed canopy of the cloud forest is still possible to find ruins of monumental architecture that remains hidden to the current affluence of tourist who fledge to see the many hummingbirds or myriad exotic flowers of this biodiversity hotspot. When tourists see *Machu Picchu* now, only few of them realize that this marvel of the world is really located amidst the most magnificent wonder of nature, the tropical montane cloud forest landscapes, that is likely to have been highly modified in ancestral time, abandoned for centuries, and now "rediscovered" by gold miners, timber companies, cattle ranchers, hungry villagers, and curious ecotourists. Where will they draw the lines in the mountains next? How many more citadels (*or "Llakta"*) we must find to accept that the ATER are hybrid cultural landscapes?

References

Allen TR, Walsh SJ, Cairns DM, Messina JP, Butler DR, Malanson GP (2004) Geostatistics and spatial analysis: characterizing form and pattern at the alpine treeline. In: Geographic information science and mountain geomorphology. Springer, Berlin, pp 189–218

Appenzeller T (2019) Global warming has made iconic Andean peak unrecognizable. Science 365(6458):1094–1097

Acosta-Solis M (1976) Vocabulario basico de fitoecologia: rectificacion terminologica de la obra Holdridge. Instituto Ecuatoriano de Ciencias Naturales. Quito, EC.

Baeid C (1999) Polylepis spp. en los Andes centrales: un análisis preliminar sobre cambios climáticos y el impacto de la actividad humana en su distribución. In: Desarrollo Sustentable de Montañas: Entendiendo las Interfaces Ecológicas para la Gestión de Paisajes Culturales en los Andes. Sarmiento F, Hidalgo J (eds) Memorias del III Simposio Internacional de la Asociación de Montañas Andinas, Quito

Bowman WD, Cairns DM, Baron JS, Seastedt TR (2002) Islands in the sky: alpine and treeline ecosystems of the Rockies. Rocky Mountain futures: an ecological perspective. Island Press, Washington, DC, pp 183–202

Bruijnzeel LA, Scatena FN, Hamilton LS (2011) Tropical montane cloud forests: science for conservation and management. Cambridge University Press, Cambridge

Bubb PI, May I, Miles L (2004) Cloud forest agenda. UNEP-WCMC biodiversity series 20. UNEP-WCMC, Cambridge

Chepstow-Lusty AJ, Bennett KD, Switsur VR, Kendall A (1996) 4000 years of human impact and vegetation change in the central Peruvian Andes—with events parallelling the Maya record? Antiquity 70(270):824–833

Churchill S, Balslev H, Forero E, Luteyn J (1995) Biodiversity and conservation of Neotropical montane forests. The New York Botanical Garden, Bronx. 702pp

Chlatante D, Scippa G, DiLorio A, Sarnataro M (2003) The influence of steep slopes on root system development. J Plant Growth Regul 4:247–260

Delibes M, Castañeda I, Fedriani JM (2017) Tree-climbing goats disperse seeds during rumination. Front Ecol Environ 15(4):222–223

Dodson CH, Gentry AH (1991) Biological extinction in western ecuador. Annals of the Missouri Botanical Garden 78(2):273

2 Dynamics of Andean Treeline Ecotones: Between Cloud Forest and Páramo... 41

Ellenberg H (1979) Man's influence on tropical mountain ecosystems in South America. J Ecol 67:401–416

Ellenberg H (1958) Wald oder Steppe? Die natürliche Pflanzendecke der Anden Perus. Umschau 1958:645–681

Elliott GP, Kipfmueller KF (2010) Multi-scale influences of slope aspect and spatial pattern on ecotonal dynamics at upper treeline in the southern Rocky Mountains, USA. Arct Antarct Alp Res 42(1):45–56

Foster P (2001) The potential negative impacts of global climate change on tropical montane cloud forests. Earth Sci Rev 55(1–2):73–106

Frascaroli F, Fjeldted T (2018) From abstractions to actions: re-embodying the religion and conservation nexus. J Stud Relig Nat Cult 11(4):511–534

Gehrig-Fasel J, Guisan A, Zimmermann NE (2007) Tree line shifts in the Swiss Alps: climate change or land abandonment? J Veg Sci 18(4):571–582

Gioda A, Hernández Z, Gonzáles E, Espejo R (1995) Fountain trees in the Canary Islands: legend and reality. Adv Hortic Sci 9(3):112–118

Hamilton L, Juvik J, Scatena F (1993) Tropical montane cloud forests. The East West Center, Honolulu. 284pp

Hamilton L, Juvik J, Scatena F (1994) The Puerto Rico tropical montane cloud forests symposium: introduction and workshop synthesis. In: Ecological studies, vol 110. Springer, New York, pp 1–23

Huisman SN, Bush MB, McMichael CN (2019) Four centuries of vegetation change in the mid-elevation Andean forests of Ecuador. Veg Hist Archaeobotany 1:1–11

Knapp G (1991) Andean ecology: adaptive dynamics in Ecuador. Westview Press, Boulder

Kapelle M, Brown A (2001) Bosques Nublados del Neotrópico. National Institute of Biodiversity (InBIO), San José. 698pp

Körner C (2012) Alpine treelines: functional ecology of the global high elevation tree limits. Springer Science & Business Media, Basel

Laimer HJ (2017) Anthropogenically induced landslides—a challenge for railway infrastructure in mountainous regions. Eng Geol 222(1):92–101

Luteyn JL, Balslev H (1992) Páramo: an Andean ecosystem under human influence. Academic Press, London

Lyons W (2019) Cloud forests of Costa Rica: ecosystems in Peril. Weatherwise 72(3):32–37

Malanson GP, Butler DR, Fagre DB, Walsh SJ, Tomback DF, Daniels LD, Resler LM, Smith WK, Weiss DJ, Peterson DL, Bunn AG (2007) Alpine treeline of western North America: linking organism-to-landscape dynamics. Phys Geogr 28(5):378–396

Malanson GP, Resler LM, Bader MY, Holtmeier FK, Butler DR, Weiss DJ, Daniels LD, Fagre DB (2011) Mountain treelines: a roadmap for research orientation. Arct Antarct Alp Res 43(2):167–177

Malanson GP, Resler LM (2016) A size-gradient hypothesis for alpine treeline ecotones. J Mt Sci 13(7):1154–1116

Martin PH, Bellingham PJ (2016) Towards integrated ecological research in tropical montane cloud forests. J Trop Ecol 32(5):345–354

Mathisen IE, Mikheeva A, Tutubalina OV, Aune S, Hofgaard A (2014) Fifty years of tree line change in the Khibiny Mountains, Russia: advantages of combined remote sensing and dendroecological approaches. Appl Veg Sci 17(1):6–16

Miehe G, Miehe S (2000) Comparative high mountain research on the treeline ecotone under human impact: carl troll's "asymmetrical zonation of the humid vegetation types of the world" of 1948 reconsidered. Erdkunde, pp 34–50

Möhl P, Mörsdorf MA, Dawes MA, Hagedorn F, Bebi P, Viglietti D, Freppaz M, Wipf S, Körner C, Thomas FM, Rixen C (2019) Twelve years of low nutrient input stimulates growth of trees and dwarf shrubs in the treeline ecotone. J Ecol 107(2):768–780

Myers N (2003) Biodiversity hotspots revisited. Bioscience 53(10):916–917

Myster RW (2012) Ecotones between forest and grasslands. Springer-Verlag, Berlin

Myster RW, Sarmiento FO (1998) Seed inputs to microsite patch recovery on two Tropandean landslides in Ecuador. Restor Ecol 6(1):1–10

Nadkarni NM, Solano R (2002) Potential effects of climate change on canopy communities in a tropical cloud forest: an experimental approach. Oecologia 131(4):580–586

Naess A (1989) Metaphysics of the treeline. Trumpeter 6(2):45

Penney D, Oschendorf J (2015) The Great Inka Road: Engineering an Empire. Smithsonian Institution.

Rahbek C, Borregaard M, Coldwell R, Dalsgaard B, Holt B, Morueta-Holme N, Nogues-Bravo D, Whottaker R, Fjeldsa J (2019) Humboldt's enigma: what causes global patterns of mountain biodiversity? Science 365:1108–1113. (special issue on Mountain Life)

Resler LM (2006) Geomorphic controls of spatial pattern and process at alpine treeline. Prof Geogr 58(2):124–138

Sarmiento FO (2016) Neotropical mountains beyond water supply: environmental services as a trifecta of sustainable mountain development. In: Greenwood G, Shroder J (eds) Mountain ice and water: investigations of the hydrologic cycle in alpine environments. Elsevier, New York, pp 309–324

Sarmiento FO (2012) Contesting Páramo: critical biogeography of the northern Andean Highlands. Kona Publishing. Higher Education Division, Charlotte, NC. 150pp

Sarmiento FO (2002) Anthropogenic change in the landscapes of highland Ecuador. Geogr Rev 92(2):213–234

Sarmiento FO (2000) Breaking mountain paradigms: ecological effects on human impacts in managed tropandean landscapes. AMBIO J Hum Environ 29(7):423–432

Sarmiento FO (1995a) Restoration of equatorial Andes: the challenge for conservation of trop-Andean landscapes. In: Churchill SH, Balslev H, Forero E, Luteyn J (eds) Biodiversity and conservation of Neotropical montane forests. The New York Botanical Garden, Bronx, pp 637–651. 702pp

Sarmiento FO (1995b) Naming and knowing an Ecuadorian landscape: A case study of the Maquipucuna Reserve. The George Wright Forum 12(1):15–22

Sarmiento FO (1987) Antología Ecológica del Ecuador: Desde la Selva hasta el Mar. Editorial Casa de la Cultura Ecuatoriana. Museo Ecuatoriano de Ciencias Naturales, Quito

Sarmiento FO, Hitchner S (2017) Indigeneity and the sacred: indigenous revival and the conservation of sacred natural sites in the Americas. Berghahn Books, New York. 266pp

Sarmiento FO, Ibarra JT, Barreau A, Pizarro JC, Rozzi R, González JA, Frolich LM (2017) Applied montology using critical biogeography in the Andes. Ann Am Assoc Geogr 107(2):416–428

Sarmiento FO, Frolich LM (2002) Andean cloud forest tree lines. Mt Res Dev 22(3):278–288

Sylvester SP, Sylvester MD, Kessler M (2014) Inaccessible ledges as refuges for the natural vegetation of the high Andes. J Veg Sci 25(5):1225–1234

Taylor BR (2010) Dark green religion: nature spirituality and the planetary future. University of California Press, Berkeley

Varela L (2008) La alta montaña del norte de los Andes: El páramo, un ecosistema antropogénico. Pirineos 163:85–95

Wang Y, Zhu H, Liang E, Camarero JJ (2016) Impact of plot shape and size on the evaluation of treeline dynamics in the Tibetan Plateau. Trees 30(4):1045–1056

White S (2013) Grass páramo as hunter-gatherer landscape. The Holocene 23(6):898–915

Young KR, Leon B (2006) Tree-line changes along the Andes: implications of spatial patterns and dynamics. Philos Trans R Soc B 362(1478):263–272

Chapter 3
Análisis Regional En Ecosistemas De Montaña En Colombia: Una mirada desde la funcionalidad del paisaje y los servicios ecosistémicos

Paola Isaacs-Cubides, Julián Díaz, and Tobias Leyva-Pinto

3.1 Introduction

Colombia tiene una superficie de 1.141.000 ha, y es altamente heterogénea en términos geográficos con cinco regiones biogeográficas que cubren una amplia gama de elevaciones (0–5800 m), tiene una gran variedad de ecosistemas, como páramos, bosque andino, bosques húmedos, bosques secos, manglares, etc.), moderada precipitación anual (300–1000 mm) y diversas características geológicas (Etter et al. 2006). Esta variabilidad ambiental en relación con el tamaño geográfico de Colombia ha resultado en altas tasas de endemismo y notable riqueza de especies (51.330 especies), convirtiendo a Colombia en el segundo país con mayor diversidad (primero en aves y orquídeas, segundo en plantas, anfibios, mariposas y peces de agua dulce, tercero en palmas y reptiles y cuarto en mamíferos; Etter et al. 2011; SIB Colombia 2019).

Esta variedad se debe en gran parte a la presencia de un sistema de cordilleras, que facilita la creación de condiciones microclimáticas que propiciaron la elevada diversificación. En estos sistemas de montaña, en buena parte se presentan bosque nublados, pero que sobretodo, albergan ecosistemas de alta montaña y páramo únicos en el mundo.

En Colombia, al menos el 40% del territorio continental está degradado (Etter et al. 2008), con una tasa de deforestación cercana a las 220.000 ha/año (IDEAM 2014). Además, la calidad y cantidad de los servicios ecosistémicos se han visto gravemente

---

The original version of this chapter was revised. The correction to this chapter is available at https://doi.org/10.1007/978-3-030-57344-7_12

P. Isaacs-Cubides (✉) · J. Díaz
Instituto de Investigación de Recursos Biológicos Alexander von Humboldt, Bogota, Colombia
e-mail: pisaacs@humboldt.org.co

T. Leyva-Pinto
Universidad Nacional de Colombia, Bogota, Colombia

© Springer Nature Switzerland AG 2021, Corrected Publication 2021
R. W. Myster (ed.), *The Andean Cloud Forest*,
https://doi.org/10.1007/978-3-030-57344-7_3

afectadas, al igual que el capital social y las relaciones entre comunidades y ecosistemas naturales (Murcia and Guariguata 2014). La continua deforestación y degradación de los ecosistemas tiene un impacto negativo en los suelos y el abastecimiento de agua, afecta a los sistemas agrícolas y otros sistemas productivos y amenaza los servicios ecosistémicos requeridos por millones de personas.

Esta degradación es el resultado de un desarrollo económico con base en la sobreexplotación de los recursos naturales y una industria extractiva, que no considera la dependencia que tiene el crecimiento económico de la capacidad que tenga el ambiente natural para tolerar todos los procesos económicos, sociales, tecnológicos y culturales (Aguilar-Garavito and Ramírez 2015).

En Colombia los ecosistemas degradados se localizan principalmente en las regiones andina y caribe, lo cual coincide con la ubicación que históricamente ha presentado la mayoría de los asentamientos humanos en el territorio (Etter and Wyngaarden 2000; MADS 2015; Ramírez et al. 2015).

En el país, alrededor de 6.000.000 ha de bosques presentan fragmentación o representan vegetación secundaria (IDEAM 2014; Ramírez et al. 2015). Esto es de gran preocupación porque después de la degradación viene la deforestación. Por otro lado, los páramos ocupan 2,9 millones de hectáreas (IAvH 2012), de las cuales un 15.9% presenta pérdida en áreas que no es permitido su uso (Isaacs 2014; Cadena-Vargas and Sarmiento 2016).

Entender estos patrones de transformación del territorio es clave para la toma de decisiones, es por ello que se requiere desarrollar diversos tipos de insumos que permitan evaluar las condiciones de intervención en el territorio.

En la actualidad se cuenta con diversas herramientas de análisis a nivel espacial, que son necesarias para determinar qué zonas son apropiadas para la conservación y cuales son prioritarias para iniciar acciones de restauración, las cuales adicionalmente facilitan la evaluación del comportamiento de las dinámicas de los ecosistemas y sus coberturas en determinada zona.

La selección de dichas áreas implica un análisis comparativo entre las diferentes variables posibles, mediante la identificación y definición de criterios apropiados para resolver y abordar el objetivo del estudio de acuerdo a las características de la zona.

Por otro lado, para evaluar cuál es el comportamiento de las coberturas en una región o paisaje, y determinar las dinámicas de ocurrencia de estos eventos, se emplean métricas o estadísticos bajo la teoría de patrones espaciales, que incorporan el paisaje representado como un mosaico de parches o coberturas discretas. Esta teoría busca explicar la distribución de los objetos geográficos, sus patrones y procesos a través del tiempo y en otras regiones (Legendre and Legendre 1998).

Evaluar estos elementos espaciales es insumo para realizar una zonificación del área y determinar las necesidades de restauración, de preservación o de uso, siendo una buena herramienta para la planificación del territorio (GREUNAL 2010).

Estas métricas se enmarcan en dos categorías generales: La composición se refiere a los atributos asociados con la variedad y abundancia de tipos de parches sin considerar el carácter espacial o ubicación de estos. La segunda categoría evalúa la configuración espacial, la cual se refiere al arreglo, posición, orientación y carácter espacial (forma, área núcleo) de los parches en el paisaje (McGarigal et al. 2012).

3.2 Methods

Adicionalmente, es necesario considerar un aspecto desde la funcionalidad del paisaje, en donde se incluyen necesidades propias de las especies como por el ejemplo la conectividad de su hábitat.

Las zonas de montaña tienen como particularidad, que la conectividad la da el gradiente de altitud, por lo que en el presente capítulo, evaluamos las condiciones de las zonas de montaña en términos de composición de bosques nublados montanos, andinos y altoandinos y su transición hacia el páramo, como un conjunto inseparable de condiciones para las especies (Fig. 3.1).

Se realizó un análisis de las condiciones de estos ecosistemas, teniendo en cuenta una cota mayor a los 1200 m que abarca hasta alturas de 5775 m, en donde se presenta gran cantidad de bosques nublados. Se evaluó la composición de coberturas usando el mapa Corine Landcover para Colombia más reciente a escala 1:100.000 y se realiza el ejercicio de conocer las diferentes condiciones del paisaje y su estado de preservación especialmente considerando el tipo de cobertura, el tamaño y la forma (relación perímetro-área) de los fragmentos de bosque, y el grado de fragmentación entre ellos (Mcgarigal et al. 2012). La obtención de la capa del índice de fragmentación se realizó a partir de la capa de uso de la tierra, en este caso se toman los sitios pertenecientes a áreas naturales los cuales se intersectan con una malla de puntos distanciados cada 300 metros. Estos puntos resultantes se procesan mediante el cálculo de densidad Kernel, a partir del cual se obtiene valores de densidad altos para zonas naturales y densidad baja para zonas transformadas (Correa-Ayram et al. 2017). Del raster obtenido, se realizó una reclasificación por cuartiles de los valores para obtener umbrales de 1 a 5, siendo 5 las áreas de mayor fragmentación por distancia entre parches. Se construyó un índice mediante la suma de los atributos, que relaciona el tamaño, la forma y el grado de fragmentación (actúa como distancia), para definir zonas de áreas naturales más degradadas (tamaños pequeños, formas pequeñas y alta fragmentación).

Adicionalmente, se realizó un ejercicio para identificar áreas con mayor valor de preservación, a partir de modelos de prestación de servicios ecosistémicos de regulación, disponible en análisis cartográficos.

Se mapearon los servicios de carbono a través del cálculo de biomasa aérea de acuerdo al tipo de cobertura siguiendo lo propuesto por Yépes y colaboradores (IDEAM 2013).

Se evaluó la oferta hídrica, a través de la interpolación de datos de estaciones climáticas para la estimación de la evapotranspiración potencial y la escorrentía (Anexo 1). Esta capa se cruzó con los tipos de cobertura, el material parental y la pendiente, para definir zonas donde hay mayor regulación hídrica por el efecto de la presencia de la vegetación, la permeabilidad y la infiltración.

Por último, se reconocieron aquellas zonas donde aún se mantiene control de erosión e inundaciones al mantener la vegetación natural, empleando los mapas de erosión nacional (2015) y susceptibilidad a inundaciones (IDEAM 2013). Del mapa

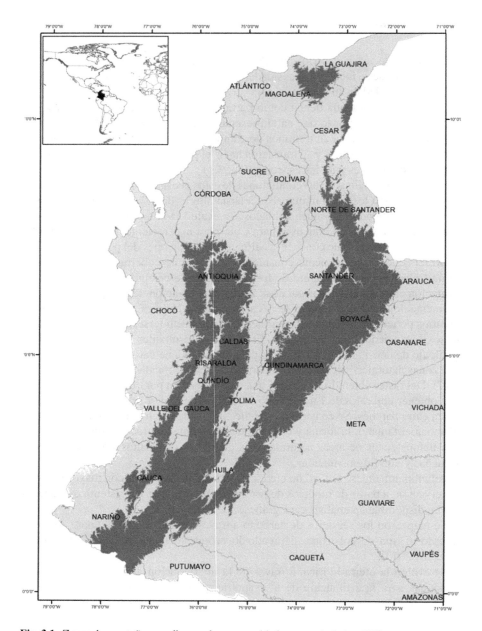

Fig. 3.1 Zonas de montaña que albergan bosques nublados, a partir de los 1200 m

de servicios, se obtuvo un mapa final acumulado, para definir zonas de importancia para la preservación.

3.3 Resultados and Discussion

De acuerdo al mapa de coberturas de la tierra, de las 18.368.518 ha evaluadas, el 45% corresponden a pastos y cultivos y el 53% a áreas naturales (Fig. 3.2).

Esto de entrada denota un estado elevado de intervención desde el inicio. También se presentan 5.900 ha de humedales de montaña, 55.000 ha de cuerpos de agua y 135.000 ha de infraestructura y ciudades (Fig. 3.3). En este rango altitudinal, se encuentra un total de 900 pueblos y ciudades lo que también evidencia la alta densidad de ocupación humana allí presente.

En cuanto a las áreas naturales presentes en ese 53%, el 65% corresponde a bosques, el 18% arbustales y el 8% pastizales naturales y vegetación secundaria en transición de crecimiento (Fig. 3.4).

De acuerdo a la distribución de los tamaños de los parches, solo existe una zona continua de coberturas naturales de más de 1.700.000 ha, hacia las zonas de Putumayo, Caquetá y Huila, al sur del país (en verde). También se destacan los bosques entre Huila y Tolima de más de 240.000 ha, los bosques entre Cauca y Valle (217.000 ha), al norte de Antioquia y noroccidente del Meta con más de 2 millones

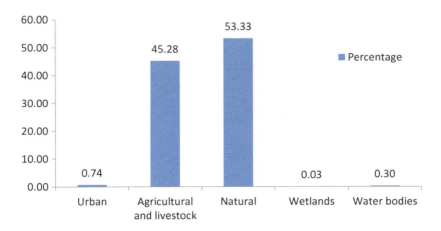

Fig. 3.2 Porcentaje de cobertura presente en las zonas de montaña

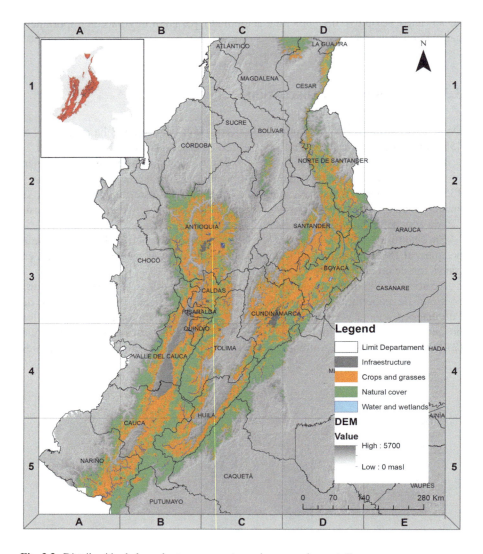

Fig. 3.3 Distribución de las coberturas presentes en las zonas de montaña

de hectáreas (en amarillo todos). Si bien aún existe una proporción alta de parches de más de 200.000 ha, se observa una gran cantidad de parches de menos de 10.000 ha, distribuidos en todo el país (Fig. 3.5).

Por su parte la relación perímetro y tamaño (métrica de forma Shape index, Mcgarigal et al. 2012), presenta una alta proporción de las formas debido al gran tamaño entre estas, con algunos parches cuyo tamaño evidencia un alto estado de

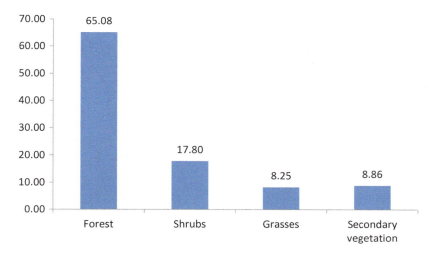

Fig. 3.4 Porcentaje de cada cobertura natural presente en las zonas de montaña

remanencia y transformación (Fig. 3.6). Esto se debe a que los parches empiezan a adquirir formas más cuadradas resultado del uso de la tierra para cultivos y ganadería y su parcelación (Mcgarigal et al. 2012).

De acuerdo a la distribución de frecuencias de los valores de forma, la mayoría se encuentra en valores de 2.7 (S.D: 1.47), seguido de valores por debajo de 1.47, que son los que corresponden a los parches más pequeños (Fig. 3.7).

Ya teniendo en cuenta la composición de las demás coberturas, al momento de evaluar el tamaño, la relación perímetro-área de las áreas naturales, y su estado de fragmentación, cómo indicador de distancia entre parches, los bosques en Antioquia, Santander y Cundinamarca, presentan mayor estado de fragmentación al también coincidir con las áreas de mayor ocupación humana, históricamente. Las zonas de la vertiente occidental de la cordillera occidental y al sur de la cordillera central y oriental, presentan mayor estado de conectividad. Sin embargo, es importante considerar que la cobertura en tierras bajas se ha perdido pero aún se mantiene la mayoría de la conectividad en la alta montaña, en especial en las zonas de páramo (Fig. 3.8).

Considerando el análisis desde los servicios ecosistémicos, el carbono es el más conocido y empleado en análisis de estimaciones de captura y almacenamiento en biomasa. En este caso, los bosques juegan el papel fundamental como elemento para cuantificar este servicio, siendo el que mayor aporte realiza. Los bosques son grandes almacenes de carbono en esta zona, por lo que la destrucción de las coberturas y el cambio de uso de la tierra representan un incremento en las emisiones de carbono como contribuyente de gases de efecto invernadero y su huella de carbono (Yepes et al. 2011). En este caso de montaña, las zonas altas pueden no ser tan representativas para este servicio, ya que los páramos por ejemplo, corresponden a coberturas de pastizales naturales, cuyo aporte de carbono en biomasa es menor

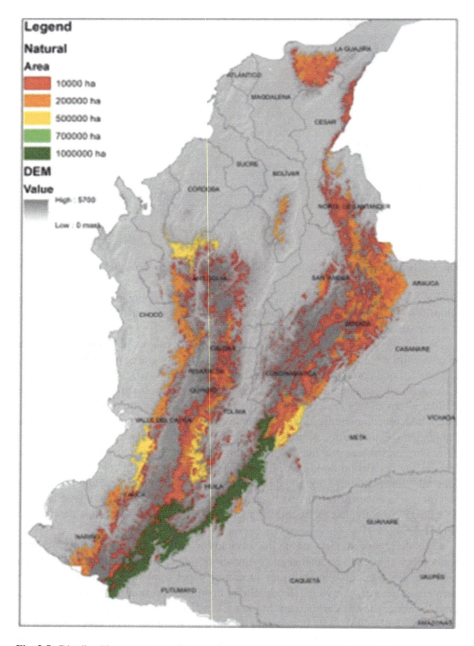

Fig. 3.5 Distribución por tamaños de las coberturas naturales

3 Análisis Regional En Ecosistemas De Montaña En Colombia...   51

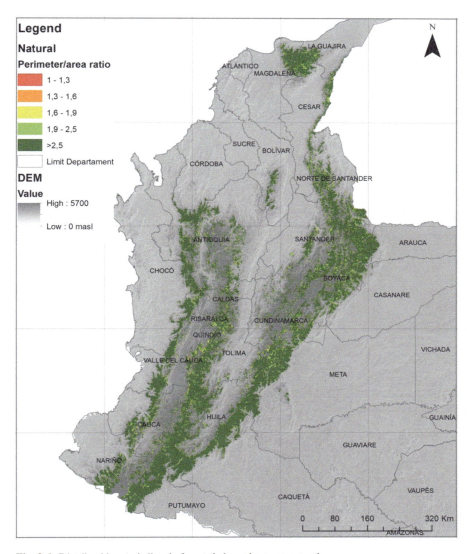

Fig. 3.6 Distribución por índice de forma de las coberturas naturales

(Fig. 3.9). Por su parte, las coberturas intervenidas presentan un menor aporte por su contenido de biomasa, pero sin incluidas en este análisis. Es por ello que empleamos otros insumos para evidenciar oferta de servicios, en especial en las áreas que aún se mantienen en pie.

Es claro que las zonas de montaña desempeñan un papel fundamental en la oferta de agua y su disponibilidad. Esto se refleja en los altos valores que pueden llegar a albergar las zonas evaluadas, con valores altos de casi 12.000 mm. Zonas del noroccidente del Meta, vertiente occidental de la cordillera occidental en Chocó, Cauca y

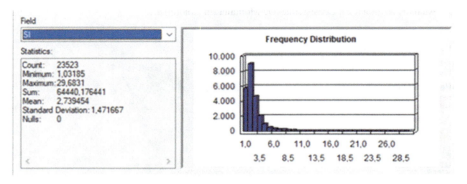

Fig. 3.7 Distribución de frecuencias de los parches naturales en la zona de montaña

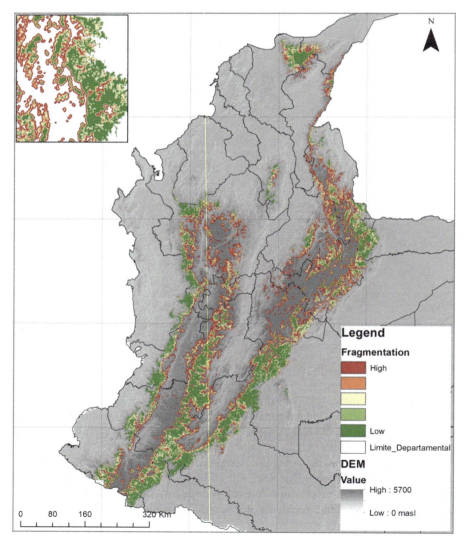

Fig. 3.8 Estado de fragmentación de las áreas naturales. En verde se observan las coberturas con mayor conectividad y en rojo, las más fragmentadas

3 Análisis Regional En Ecosistemas De Montaña En Colombia…

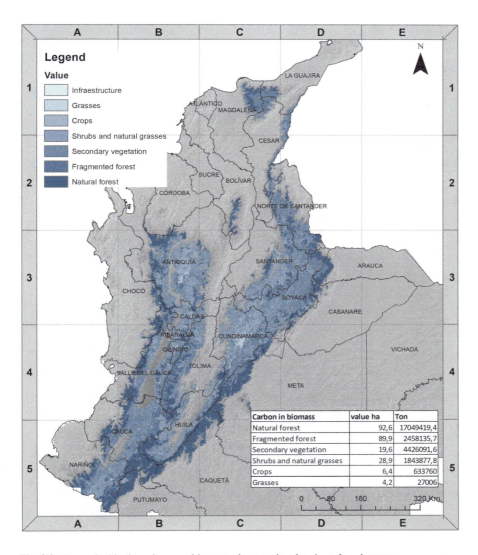

Fig. 3.9 Acumulación de carbono en biomasa, de acuerdo a los tipos de cobertura

Nariño, cordillera central en Antioquia, son lugares de elevada importancia por la acumulación de agua, en especial aquella que viene de las mayores altitudes (Fig. 3.10).

Una vez incluido el tipo de cobertura, la pendiente y el material parental, es posible evidenciar las zonas donde hay mayor regulación de esa oferta de agua, al mantener la vegetación natural aún en pie. En especial las zonas al pie de la montaña son las que albergan mayor cantidad de servicios (Fig. 3.11).

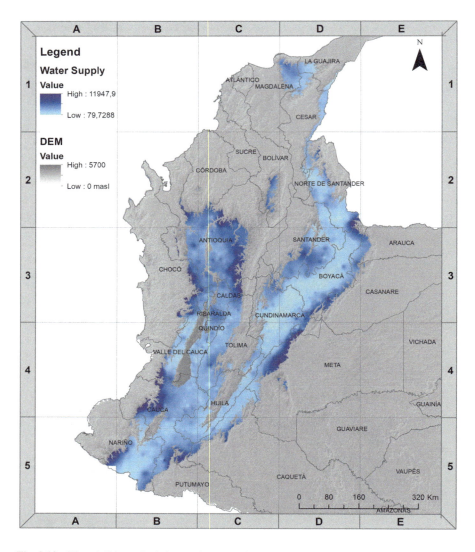

Fig. 3.10 Oferta hídrica calculada para las zonas de montaña

Es importante destacar, que para las zonas de montaña adicionalmente el servicio de control de erosión es de gran preponderancia, ya que actualmente la zona se ve muy afectada por eventos de remoción en masa por el mal manejo del suelo (Servicio Geológico Colombiano 2013). En estas zonas de montaña, se presenta control de erosión en 1.688.375 ha, sin embargo existe una alta cantidad de zonas ganaderas, lo que aumenta aún más la susceptibilidad de erosión (Fig. 3.12).

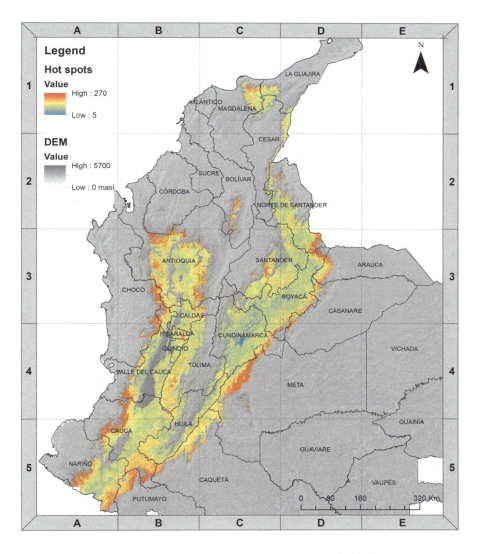

Fig. 3.11 Mapa de hotspots de servicios de carbono, oferta y regulación hídrica

La modelación de los servicios ecosistémicos permite darle valor monetario, a un valor casi intangible que poseen las áreas naturales y esto ha podido ser traducido a cifras. Por nombrar algunos, de acuerdo con Petley (Dinero 2017), cerca de 100 colombianos mueren al año a causa de deslizamientos de tierra, siendo el 50% de las muertes asociadas a desastres naturales, seguida de las inundaciones (42%) y los

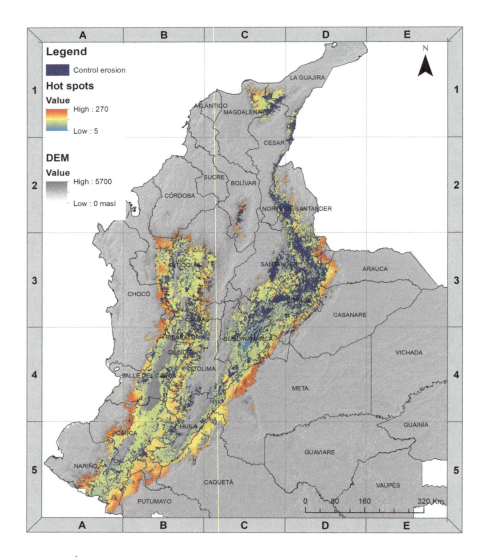

Fig. 3.12 Áreas de control de erosión en coberturas vegetales en la montaña

terremotos (8%). Asimismo, en términos de costos por ejemplo, en el año 2017 ocurrió una gran avalancha en la ciudad de Mocoa, la cual tuvo un costo inicial de cerca de 13 millones de dólares para sólo atender la emergencia (Dinero 2017), con afectación a cerca de 10.000 viviendas que aproximadamente tienen un costo de 27 millones de dólares para su reemplazamiento. Las donaciones nacionales llegaron a cerca de 3.5 millones, mientras que las internacionales ya alcanzaron los 9 millones, según cifras de la Unidad Nacional de la Gestión del Riesgo (Razón pública 2018).

Según el Plan Nacional de Desarrollo 2010–2014, hubo 269 acueductos y 751 vías afectadas, se estimó que murieron 600.000 aves y 115.000 bovinos, aparte.

del desplazamiento de 1.430.200 animales y la pérdida de 2.601 toneladas de carne, así como la pérdida de 1.080.000 ha de área cultivable (Sánchez 2011). Por su parte, la atención de la emergencia invernal fue estimada en $8.5 billones de dólares a precios en 2010, además de los sobrecostos que se generaron por la oferta de alimento en las poblaciones que se abastecen de productos de la región (Sánchez 2011). Estos costos pudieron ser evitados o invertidos en una adecuada gestión del territorio, con miras a la prevención y mitigación de los desastres.

Sin tener en cuenta el valor intrínseco de las áreas naturales, el mal manejo del suelo y su depreciación en términos de servicios ecosistémicos acarrea costos humanos y económicos altos. Su rehabilitación podría disminuir estos gastos y generar economías más rentables con una mejor productividad.

En este sentido la presencia de áreas protegidas es una de las opciones que los gobiernos adoptan para garantizar la preservación de la biodiversidad, pero también para garantizar la prestación de servicios. Para las zonas de montaña se han realizado grandes esfuerzos para aumentar la declaratoria de áreas de preservación (Fig. 3.13). En su mayoría estas áreas buscan proteger ecosistemas de páramos, por su oferta de agua, pero también ha aumentado la representatividad de estas áreas en tierras más bajas.

6.264.498 ha se encuentran en estado de protección, sin embargo es importante garantizar su real preservación, teniendo en cuenta condiciones sociales desfavorables en Colombia y poco acceso a la tierra. Es necesario proponer y apoyar tipos de uso más acordes con la capacidad productiva de los suelos, las cuales mejoren las condiciones de vida de los habitantes. Muchas de las zonas modeladas y que contienen alta oferta se servicios ecosistémicos, cómo elevada integridad en sus áreas naturales, aún están por fuera de alguna categoría de protección.

Así mismo, se debe mejorar en el establecimiento de áreas para la conectividad entre estas áreas protegidas, las cuales en su mayoría están separadas. Esta conectividad entre páramos y la alta montaña, es clave para seguir manteniendo las condiciones naturales. Dichas estrategias pueden incluir sistemas de producción hacía una transición similar a la de los bosques y sus productos derivados.

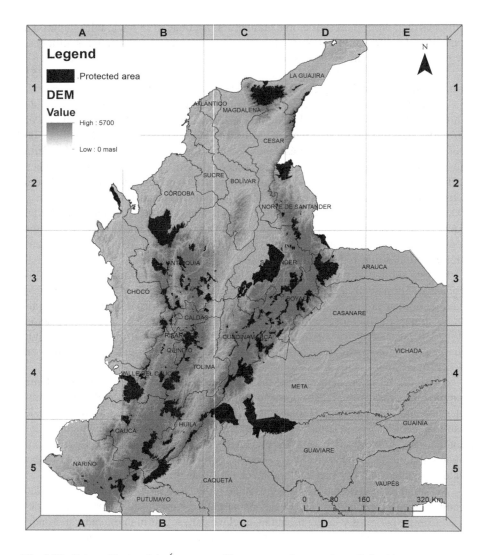

Fig. 3.13 Sistema Nacional de Áreas protegidas en zonas de montaña en Colombia

References

Aguilar-Garavito M, Ramírez W (2015) Monitoreo a procesos de restauración ecológica, aplicado a ecosistemas terrestres. Instituto de Investigación de Recursos Biológicos Alexander von Humboldt (IAvH), Bogotá D.C., 250 pp

Camilo A. Correa A, Manuel E. Mendoza, Andrés E, Diego R. Pérez S (2017) Anthropogenic impact on habitat connectivity: A multidimensional human footprint index evaluated in a highly biodiverse landscape of Mexico. Ecological Indicators 72:895–909

Cadena-Vargas CE, y Sarmiento CE (2016) Cambios en las coberturas paramunas. En Gómez, MF, Moreno, LA, Andrade GI, y Rueda C (Eds.). Biodiversidad 2015. Estado y tendencias de

3 Análisis Regional En Ecosistemas De Montaña En Colombia...   59

la biodiversidad continental de Colombia. Instituto Alexander von Humboldt. Bogotá, D.C., Colombia

Dinero (2017) https://www.dinero.com/empresas/confidencias-on-line/articulo/tragedia-de-mocoa-tiene-un-costo-no-calculado/243678

Etter A, van Wyngaarden W (2000) Patterns of landscape transformation in Colombia, with emphasis in the Andean region. Ambio 29:432–439. https://doi.org/10.1579/0044-7447-29.7.432

Etter A, McAlpine C, Wilson K, Phinn S & Possingham H (2006) Regional patterns of agricultural land use and deforestation in Colombia. Agricultural Ecosystems Environment 114, 369–386. https://doi.org/10.1016/j.agee.2005.11.013

Etter A, McAlpine C, Possingham H (2008) Historical patterns and drivers of landscape change in Colombia since 1500: a regionalized spatial approach. Ann Assoc Am Geogr 98:2–23. https://doi.org/10.1080/00045600701733911

Etter A, McAlpine C, Seabrook L, Wilson K (2011) Incorporating temporality and biophysical vulnerability to quantify the human spatial footprint on ecosystems. Biol Conserv 144:1585–1594. https://doi.org/10.1016/j.biocon.2011.02.004

GREUNAL (Grupo De Restauración Ecológica) (2010) Guías técnicas para la restauración ecológica de ecosistemas. Convenio de asociación no. 22 entre Ministerio de Ambiente, Vivienda y Desarrollo Territorial (MAVDT) y la Academia de Ciencias Exactas, Físicas y Naturales (ACCEFYN). Departamento de Biología, Facultad de Ciencias, Universidad Nacional de Colombia

IAvH (2012) Cartografía de Páramos de Colombia Esc. 1:100.000. Proyecto: Actualización del Atlas de Páramos de Colombia. Convenio Interadministrativo de Asociación 11–103, Instituto Humboldt y Ministerio de Ambiente y Desarrollo Sostenible. Bogotá D.C. Colombia

IDEAM (2013) Mapa de susceptibilidad de inundación escala 1:100.000

IDEAM (2014) Subdirección de Ecosistemas e Información Ambiental. Grupo de Bosques. Proyecto Sistema de Monitoreo de Bosques y Carbono. Bogotá, D.C., Colombia

Isaacs P (2014) Composición y configuración de los páramos de Colombia. En: Restauración ecológica de los páramos de Colombia. Transformación y herramientas para su conservación. Instituto de Investigación de Recursos Biológicos Alexander von Humboldt (IAvH), Bogota, D.C., 296 pp

Legendre P, Legendre L (1998) Numerical ecology. Elsevier, Ámsterdam

McGarigal K, Cushman SA, Neel MC, Ene E (2012) FRAGSTATS: spatial pattern analysis program for categorical maps. Computer software program produced by the authors at the University of Massachusetts, Amherst. www.umass.edu/landeco/research/fragstats/fragstats.html

Ministerio de Ambiente y Desarrollo Sostenible de Colombia (MADS) (2015) Plan Nacional de Restauración: restauración ecológica, rehabilitación y recuperación de áreas disturbadas. Bogotá, D.C., Colombia. ISBN: 978-958-8901-02-2

Murcia C, Guariguata MR (2014) La restauración ecológica en Colombia: Tendencias, necesidades y oportunidades. Occasional paper 107. CIFOR, Bogor. https://doi.org/10.17528/cifor/004519

Ramírez W, Murcia C, Guariguata M, Thomas E, Aguilar M, Isaacs-Cubides P (2015) Restauración Ecológica, los retos para Colombia. En: Biodiversidad 2015. Estado y tendencias de la biodiversidad continental De Colombia. Instituto de Investigación de Recursos Biológicos Alexander von Humboldt (IAvH), Bogota, D.C.

Razón pública (2018) Los costos de las lluvias torrenciales: el caso de Mocoa. https://www.razonpublica.com/index.php/econom-y-sociedad-temas-29/11345-los-costos-de-las-lluvias-torrenciales-el-caso-de-mocoa.html

Sánchez (2011) Documentos de trabajo sobre economía regional. Banco de la República, Centro de Estudios Económicos Regionales. ISSN 1692–3715 https://www.banrep.gov.co/sites/default/files/publicaciones/archivos/DTSER_150_0.pdf

Servicio Geológico Colombiano (2013) Mapa de susceptibilidad de erosión escala 1:100.000

SIB Colombia (2019) https://sibcolombia.net/biodiversidad-en-cifras-2019/

Yepes AP, Navarrete D, Duque aJ, Phillips JF, Cabrera KR, Alvarez E, García MC, Ordoñez MF (2011) Protocolo para la estimación nacional y subnacional de biomasa - carbono en Colombia

Chapter 4
Ecohydrology of Tropical Andean Cloud Forests

Conrado Tobón

4.1 Introduction

Montane ecosystems around the world are found from the equator to the poles and occupy approximately one fifth of the surface of continents and islands (Ives et al. 1997). In South America, the Andes, as the longest ridge in America, extends over approximately 1.5 million km^2, running from 11° N to 23° S, with altitudes up to 6000 masl. The tropical part runs mainly from Venezuela through Colombia, Ecuador, and Peru, comprising area of approximately 35,824 km^2 (Cuesta et al. 2009). Main ecosystems in these environments are the glaciers, páramos, and montane forests, including cloud forests; however, montane forests cover most of the region, whereas páramos are insular formations around the highest peaks (Smith and Cleef 1988).

The Tropical Andean Cloud Forests (TACF) are high elevation forests that appear on the Andean ridge flanks, between 2000 and 3200 masl, altitudes that change according to the specific site exposition, thus cloud or fog presence (Bruijnzeel et al. 2011), and it is characterized by middle size trees (around 12 m high), abundant epiphytes hanging from tree trunks and branches, which are embedded frequently in fog, and low intensity and long duration rainfall events (Tobón 2009). Associated with the altitude and fog conditions, the temperature is low, air humidity is high, and solar radiation is patchy and predominantly low.

Several attributes make these ecosystems of special importance as the high biological diversity (Bruijnzeel et al. 2010), with endemic plants (Sklenár and Ramsay 2001; Myers et al. 2000), large water supply (Tobón 2009), and dominant climate conditions of high air humidity and low temperatures. Consequently Andean ecosystems are important not only for local people, but for the millions living in the mid and lowlands, as main rivers bring fresh water from those ecosystems, which,

---

C. Tobón (✉)
Universidad Nacional de Colombia, Medellín, Colombia
e-mail: ctobonm@unal.edu.co

© Springer Nature Switzerland AG 2021
R. W. Myster (ed.), *The Andean Cloud Forest*,
https://doi.org/10.1007/978-3-030-57344-7_4

through the vegetation, including the bryophytes hanging from trees and standing on soil surface (Tobón et al. 2010a), capture moisture from air masses and together with rainfall falling throughout the year, infiltrates into the soil, and gets further available as baseflow (Tobón 2009), during least rainfall periods.

Ecohydrology became the emergent science dealing with the study of the relationships and mechanisms among climate, soils, and vegetation with the hydrological processes at different scales, and the dynamics beyond these interactions seems to differ between ecosystems (Rodríguez-Iturbe 2000). The ecohydrological significance of cloud forest ecosystems lies on the interaction between the specific climate conditions, which includes the presence and frequency of fog and low clouds, the vegetation, including epiphytes (Bruijnzeel et al. 2010), and the reach organic matter soils (Tobón et al. 2010b), playing an important role on water infiltration and storage capacity, thus on water regulation. The provision and regulation of hydrological services of these ecosystems are explained by a combination of combined inputs of water, through rainfall and fog deposition, low evapotranspiration, the high infiltration capacity of litter and soils, and moss and soil water storage (Tobón 2009). Overall, these processes need to be characterized in TACF, and provide a unique opportunity to understand the hydrological processes linked to the ecology of Andean cloud forests.

Worldwide native forest is changing rapidly (Condit et al. 2005), and Andean cloud forests are not the exception (Wassenaar et al. 2007; Bruinsma 2003). Contradictory, while growing population demands larger amounts of ecosystems services, as water, conversion of TACF to crop plantation and pastures affects, at different rates, ecosystem water yield and water regulation (Bruijnzeel et al. 2011). Moreover, these ecosystems are also particularly vulnerable to climate change (Pounds et al. 1999; Barradas et al. 2010), thus the potential loss of actual specific ecohydrological functioning. Overall, under the contemporary conditions of widespread land use changes and climate change, water supply to population depending on water from Andean cloud forest is at risk.

Recently there has been much concern about the hydrological importance of paramo ecosystems, under the consideration that they provide water for cities in Andean region (Tobón 2009; Bruijnzeel 2004). This has enhanced by those authors that consider that disappearance of Andean glaciers, as it is happening at high rates in the last decades, will affect water supply for population in the region (Kaser et al. 2005). All this concern left out the Andean cloud forests, which may be as important, or even more important, for water supply that the previous ecosystems (Tobón 2009) considering that the extension of the area occupy by the TACF is much larger than the two foregoing ecosystems, and annual rainfall amounts are also higher (Tobón 2009).

Although the tropical ecosystems offer a great opportunity to understand the hydrological processes linked to the ecology of the native and transformed ecosystems, the number of ecohydrological research projects remain limited. This was demonstrated in a recent paper, which shows that the number of ecohydrological studies carried out in the tropics remain very low, as compared from nontropical regions (Wright et al. 2017). The number of these ecohydrological studies decrease considerably for the Neotropics (Bendix et al. 2008), and in spite of the ecological and hydrological importance, the studies of the ecohydrology of the Andean cloud forests have received only marginal attention (Garreaud 2009).

Interest in understanding ecohydrological processes of forest ecosystems and the role of forests in affecting water supply and ecosystem services has been triggered by recent worldwide water crisis and the ongoing climate change debate. Although some processes governing the ecohydrological functioning of ecosystems, as transpiration, rainfall interception, groundwater flows were early understood (e.g. Hursch and Brater 1941) and remains up today (Bruijnzeel et al. 2011), other variables as fog water inputs and soil water dynamics have not been widely considered. Moreover, the actual increased pressure on water resources, as water supply to increase planet population, but also for crop production demanded (Jackson et al. 2001), makes ecohydrology a timely science.

It should be stated here that in this century, among several papers, some comprehensive books have been published concerning cloud forest (Bruijnzeel and Hamilton 2000; Nadkarni and Wheelwright 2000; Bruijnzeel 2001; Bruijnzeel et al. 2010; Hamilton et al. 1995), and some of those included chapters regarding cloud forest hydrology (Bruijnzeel et al. 2010), however, little information was included in those, specifically related to Tropical Andean cloud forest, which agrees with the scarce ecohydrological investigations carried out in these ecosystems in last decades.

Understanding the complex ecohydrological functioning of TACF and their role in providing hydrological services is a key issue for policy makers, institutions focused on conservation, and restoration of millennium ecosystems providing environmental services, and for population depending on water flowing from these ecosystems. Therefore, the objective in this paper is to present the current understanding of the ecohydrological functioning of Tropical Andean Cloud Forests, through determining and discussing the critical parameters that control this functioning. It is also proposed the future research needed for more comprehensive understanding of ecohydrology of these ecosystems, and the water supply regimes.

4.2   Methodology

In this paper we present the current understanding of the ecohydrological functioning of Tropical Andean Cloud Forest, which is, results from ecohydrological studies made in cloud forests in Venezuela, Colombia, Ecuador, and Peru. The results presented in this paper come from two different sources: the revision of peer-review articles in English and Spanish about ecohydrology and/or hydrology of Andean cloud forests published in Science Direct, ISI Web of Science, Scopus, Redalyc, Google Scholar, and SciELO. Database was scrutinized through 2019. Some documents of a more local nature have also been reviewed, such as thesis and project reports (see Table 4.2). Further within the revised papers, we search for cited literature focused on hydrology or ecohydrology of the TACF, including Master and PhD thesis.

The second source of data consisted in unpublished information generated by the research group "Hydrology and modelling of ecosystems, Universidad Nacional de

Table 4.1 Location and main characteristics of TACF studied in Colombia

Site 1. Cloud Forest Pantano Redondo—Zipaquirá

Location 5° 02′30.30″ N 74° 02′02.19″W	Altitude 3160 masl
Average annual rainfall 1615 mm	Average temperature: diurnal 14.7 °C and night 7,7 °C
Exposition: Exposed to the Cundiboyacense Montane Plateau, with air masses coming from the dry savannah	
Main soil classes (Soil Survey Staff 2004)	Humitropets, Dystrandepts, Cryumbrepts e Histosoles Typic Placudands, Typic Hapludands, Lithic Hapludands, Pachic Melanudands y Typic Endoaquand (USDA 2014)
Main plant species	*Weinmannia tomentosa* (encenillo), *Drimys granadensis* (canelo de páramo), *Clusia multiflora* (gaque), *Hedyosmum bonplandianum* (granizo), *Cervantesia* (santalaceas), *Ilex* (acebos), *Vallea* (raques), *Escallonia* (tobos) y *Myrica* (laurel de cera) (Gentry 1991; CAR 2001)
Predominant land use	Cloud forest, surrounded by pastures and crops

Site 2. Parque Nacional Natural Chingaza, Mundo Nuevo

Location 4° 40′13.13″ N 73° 50′36.01″W	Altitude 3040 masl
Average annual rainfall 3280 mm	Average temperature. Diurnal 13.2 °C, night 9.3 °C
Exposition: Exposed to the East, with air masses coming from the Amazonia and Orinoquia (Colombia)	
Main soil classes (Soil Survey Staff 2004)	Typic Hapludands, Humic Dystrudepts, Humic Lithic Dystrudepts, Typic Humudepts, Lithic Hapludands, Pachic Melanudands y Lithic Melanudands (USDA 2014)
Main plant species	*Weinmannia spp., Weinmannia microphylla, Drimys granadensis, Clusia* cf. *Multiflora y Lepechinia conferta* (este estudio)
Predominant land use	Cloud forest and some small areas with pastures

Site 3. Cloud Forest Estrella de agua—Salento

Location 4° 37′17.59″ N 73° 25′44.73″ W	Altitude 3210 masl
Average annual rainfall 2164 mm	Average temperature. Diurnal 12.6 °C and night 7,2 °C
Exposition: Exposed to the West, air masses coming from the Pacific	
Main soil classes (Soil Survey Staff 2004)	Andic Humicryepts, Typic Melanocryands, Typic Humudepts y Typic Hapludands. (IGAC 2004)
Main plant species	Ceroxylon quindiuense, Poulsenia armata, Ficus sp., Clusia lineata, Clusia sp., Lippia hirsuta, Hyeronima oblonga y Cuatresia riparia (este estudio)
Predominant land use	Undisturbed cloud forest

Colombia," through the research projects developed in three Andean cloud forest in Colombia, since 2007 (Table 4.1). The climate of each site was characterized by installing an automatic weather station (CampbellSci Ltd) to measure variables as precipitation, temperature, solar radiation, air humidity, wind speed, and direction.

Table 4.2 The magnitude of the ecohydrological variables in Tropical Andean cloud forests (NI = No Information). *F* means fog inputs measured with a cylinder and *F*2 is fog inputs measured as net precipitation under forest, only during fog events

Authors	Location (coordinates and altitude)	Precipitation (P) and fog inputs (F) (mm y^{-1})	Temperature (°C) and relative air humidity (%)	Evapotranspiration forest interception (mm y^{-1})	Average soil water content
This study—site 1 (Tobón and Arroyave 2007)	Colombia 5° 04′56.02″ N 74° 06′12″ W 3074 masl	$P = 1615$ $F = 334$ $F2 = 65$	14.7 83.5	434 511	0,58 cm^3·cm^{-3}
This study—Site 2	Colombia 4° 40′13.13″ N 73° 50′36.01″W 3040 masl	$P = 3280$ $F = 465$ $F2 = 154$	13.2 89.0	389 575	0,63 cm^3·cm^{-3}
This study—site 3	Colombia 4° 37′17.59″ N 73° 25′44.73″ W 3210 masl	$P = 2164$ $F = 421$ $F2 = 128$	12.6 91.0	407 485	0,65 cm^3·cm^{-3}
Ramírez et al. (2017)	Colombia 72.900 E 5.243 N 2048 masl	$P = 4588$ $F = 80$	14.5 96.3	NI NI	46%
Vásquez (2016)	Colombia 4° 49′ 10″ N, 75° 33′ 50″ W 2000 masl	$P = 2472$ $F = $ ND	15.0 85.0	929 NI	35%
Burbano-Garcés et al. (2014)	Colombia 2° 29′ N and 76° 32′ W 1870 masl	$P = 1488$ $F = $ ND	20 72	NI 256	NI
Clark et al. (2014)	Peru 13° 3′ 37″ S, 71° 32′ 40″ W 2805 masl	$P = 3112$ $F = 316$	14.2 NI	688 226	NI
Jarvis and Mulligan (2010)	Colombia Andes 2° 21′ 47″ N and 76° 24′ 28″ W 2000–2600 masl	$P = 2000$ $F = $ NI	17.7 NI	NI	NI

(continued)

Table 4.2 (continued)

Authors	Location (coordinates and altitude)	Precipitation (P) and fog inputs (F) (mm y^{-1})	Temperature (°C) and relative air humidity (%)	Evapotranspiration forest interception (mm y^{-1})	Average soil water content
Oesker et al. (2010)	Ecuador 3° 58′ S and 79° 04′ W 2275 masl	P = 2737 F = NI	16.2 NI	NI 247	NI
León Peláez et al. (2010)	Colombia 06° 18″ N 75° 30″ W 2490 masl	P = 1725 F = NI	15.2 84.0	NI 250	46%[a]
Bendix et al. (2008)	Ecuador 3° 58′ 30″ S and 79° 4′ 25″ W 2270 masl	P = 2193 F = 210	15.8	570 1010	44.7%[b]
	Ecuador 2660 masl	P = 4779 F = 527	13.4 86.3		
	Ecuador 04° 06′ 71″ S, 79° 10′ 581″ W 3180 masl	P = 4743 F = 1958	9.4 93		49.1%[b]
Gómez-Peralta et al. (2008)	Peru 10° 31′ 47″ S, 75° 21′ 23″ W 2815 masl	P = 2753 F = 221	14.5 ND	NI 211	NI
Garcia (2007)	Colombia 4° 50′ N and 75° 30′ W 2550 masl 3370 masl	P at 3370 = 1453 F = 0	12.8 95	NI 265	NI
Fleischbein et al. (2006, 2005)	Ecuador 4° 00′ S and 79° 12′ W 2000 masl	P = 2592 F = ND	15.2 NI	471 985	NI
	Colombia 04° 30′ 42″ N 74° 01′ 41″ O 3000 masl	P = 1313 F = ND	13.2	465 685	70%

4 Ecohydrology of Tropical Andean Cloud Forests

Fonseca and Ataroff (2005)	Colombia 5° 25′ 13″ N, 75° 45′ 17″ W 2350 masl	$P = 3153$ $F = 438$	14.5 91	NI 1580	NI
Ataroff (2002, 2005)	Venezuela – La Mucuy 8° 38′ N 70° 02′ W 2300 masl	$P = 3125$ $F = 300$	14.0	498 1751	26%
Ataroff and Rada (2000)	Venezuela (8° 38′ N and 71° 02′ W) 2350 masl	$P = 3124$ $F = 309$	14.0 90.3	558 1687	23%
González (2000)	Colombia 2° 30′ N, 76° 60′ W 2050 masl	$P = 4120$ $F = 371$	13.0 94.0	NI NI	NI
Jarvis (2000)	Colombia 2.30 N, 76.59 E 1525 masl	$P = 3900$ $F = 2854$	19.0 90.0	NI NI	NI
Ataroff (1998)	Venezuela 08° 36′ 59.95″ S, 71° 03′ 00.31″ W 2300 masl	$P = 2959$ $F = 91$	12.0 NI	NI 1331	NI
Veneklaas and Van Ek (1990)	Colombia 4° 50′ N and 75° 30′ W. 2550 masl	$P = 2115$ $F = 0$	15.4 91	NI 262	NI
Cavalier and Goldstein (1989)	Venezuela 2500 masl	$P = 1983$ $F = 72$	8.0 NI	NI NI	NI
Steinhardt (1979)	Venezuela 2300 masl	$P = 1575$ $F = $ NI	NI NI	675 305	NI

[a]Tobón (2009)
[b]Moser et al. (2008)

Total and average values were registered each 15 min. At each site, rainfall was also measured nearby the forests, but outside the forest, by installing two automatic rain gauges (Texas instruments) at different sites, which were programmed to record total rainfall each 15 min. To measure fog inputs, we used Juvik-type fog gauges, consisting of a louvered cylindrical aluminum shade screen (Juvik and Ekern 1978). The fog gauges have a diameter of 12.8 cm and 42 cm in height, with a conical aluminum cover on top, of 65 cm diameter to minimize the contribution of rainfall. To quantify inputs, fog gauges were installed at 1.5 m height and connected to a tipping bucket rain gauge (Texas instruments), recording data every 15 min. Fog inputs were also determined through net precipitation measurements, which is further explained.

At each cloud forest, two plots of 20x50 m were randomly selected inside each forest site, to measure throughfall. This was measured by installing 4 V-shaped stainless steel throughfall gutters of 4 m length and 0.32 m width, at each plot. These troughs were placed at inclination of about 15° angle to facilitate drainage of the collected water into a pre-calibrated tipping bucket of 51 mL (Vrije Universiteit Amsterdam), which gives a resolution of about 0.051 mm per tip. Each tipping bucket was connected to Tinytag data logger (Gemini data logger), which measured water inputs each minute and register total values each 15 min, coupled with rain gauges outside the forest. To characterize the effects of forest structure on throughfall, gutters were randomly moved to a new site (within the 20 × 50m), each 3 months. Troughs were also used to measure water inputs generated by drips falling from the canopy during fog events. In this case, we consider fog inputs those values registered, either in isolated fog events (events without rainfall) or the dripping occurring sometime after rainfall had ceased when normally fog comes after the rain (based on field observations inside the forests, for the studied sites in Colombia, we consider a lapse of 15 min, as the time required for the forests to drain all drops from the rain). For better precision on fog inputs, we also calculated the amount of water required to wet the inner part of the troughs, allowing us for the real determination of water inputs in fog events, but also for throughfall, therefore a value of 0.4 mm was added to all data registered, either to throughfall or to fog.

To determine the wetting of leaves during rainfall and fog events, leaf wetness was measured using a Decagon dielectric wetness sensor (LWS-L Decagon Devices), connected to a CR1000 datalogger (Campbell Sci). LWS were up-facing installed at three different heights, from the tree crown: one on top, the second 2 m below, and the third one, at 5 m below the treetop. Sensor outputs are mV, where the higher the mV registered on the datalogger, the larger the leaf wetness. Measurements were made at intervals of 30 s with averages each 15 min. Sensors were calibrated at the laboratory, to interpret, first the value at which the sensor, as the canopy, is dry, second, the rate at which canopy is getting wet, and third, the percentage of the canopy that got wet. To this purpose, before taking the sensors to the field, in the laboratory we wet part by part the sensors (from around 5% till 100%), and made visual observations of the fraction of the sensor which was wet, and compare with measurements (mV) registered in the datalogger, at time intervals of 10 s.

The second component of net precipitation is stemflow. Some authors indicate that in cloud forests the amounts of stemflow are insignificant (<2% of gross precipitation) when compared to throughfall (Bruijnzeel 2004; Dietz et al. 2006; Gómez-Peralta et al. 2008; Holwerda et al. 2010; Bruijnzeel et al. 2011; Muñoz-Villers et al. 2012), however, in studied cloud forest in Colombia, stemflow amounts were measured at each plot and each forest, according to Tobón et al. (2000). Therefore amounts of water flowing down the tree trunks were measured using flexible transparent tubing that was cut longitudinally and wrapped in a downward spiral around tree trunks. The tubing was nailed to the tree trunk, and sealed with silicone sealant. Tubing were channeled into a tipping-bucket guage (Texas instruments), connected to a Tinytag data logger (Gemini dataloggers). Stemflow was computed (mm), by using the projected area of each tree, as the input exposed area to rain. Net precipitation was computed as the sum of throughfall plus stemflow for each rainfall event, and rainfall interception was calculated as the difference between gross rainfall outside the forests and net precipitation.

At each specific site (plots of 20x50m), volumetric soil water content (θ, cm^3 cm^{-3}) was automatically measured. To this purpose, we used Time Domain Reflectometry (TDR) sensors connected to a CR1000 datalogger (Campbell Scientific Ltd., Shepshed, UK). At each site, a soil pit of 2.0 m × 1.5 m × 1.5 m was excavated and TDR sensors were horizontally installed at five different depths representing the respective soil horizons (5, 15, 35, 50, 80, or 100 cm depth). The probes were dug into the upslope face of the pit with the depth of the hole being equal to the measuring depth to minimize the effect of soil disturbance created by digging into the soil to install sensors. The pits were backfilled after installing the TDR probes, making sure to return the respective soil layers to their original positions. Measurements were made on a minute basis, registering average values each 15 min, over the total measured period at each forest (Table 4.1). Field measurements were calibrated according to Tobón et al. (2010b).

For most investigated sites in TACF no discharge was measured, except for two out of the three sites in Colombia and sites in Ecuador (Bendix et al. 2008). Consequently water yield from these ecosystems was calculated using the water balance approach, applying the general equation $Q = P + F - ETa - I - dS/dt$, where Q is discharge, P is rainfall, F is the net water input by fog, I is the forest interception of rainfall, and dS/dt is the storage changes per time step, all in units of mm y^{-1}.

4.3 Results and Discussion

4.3.1 The Ecohydrology of TACF

The Andes mountain range is the result of the tectonic activity and continental drift in the South American continent (Brown and Lomolino 1998; Pielou 1979), which originated from the collision of the Nazca plate with the South American plate

(Fittkau et al. 1968), event that gave rise to a complex chain of mountains that extends from the south of the continent to Venezuela and includes a variety of ecosystems, including the tropical Andean cloud forests (TACF). Although TACF do not have a defined altitude range, as they are distributed at different altitudes depending on the environmental conditions existing at each site, the conditions of temperature, mainly dependent on the altitude, their exposure to currents of humidified air masses, and the condition of inversion of the trade winds (Stadtmüller 1987), in South America these ecosystems are mostly located above 2000 and below 3200 masl., that is, below the limit of the páramos, where those exists (Beck et al. 2008; Young 2006; Rada 2002; Fontúrbel 2002; Föster 2001; Hamilton et al. 1995; Stadtmüller 1987), and main differences between them, concern to the amount and frequency of fog (Tobón 2009; Lawton et al. 2001), which normally can be seeing by the amounts of mosses on tree branches and trunks.

Unlike other Andean forests, the Tropical Andean cloud forests have specific vegetation composition, that is, the abundance of epiphytes, mainly mosses, liverworts, lichens, bromeliads, and orchids attached to the tree trunks or hanging from the branches (Fig. 4.1), which largely constitute the lower strata or undergrowth of these ecosystems. While some authors reported biomass values of around 16 ton/h for Costa Rican cloud forests (Köhler et al. 2007), Hofstede et al. (1993) reported a value of 44 ton/h for TACF in Colombia. Values may range from site to site (Köhler

Fig. 4.1 Tropical Andean Cloud forest with epiphytes hanging from tree trunk and branches

et al. 2011), which seems to be dependent on the permanence and frequency of fog, among other variables, as disturbance (Fig. 4.1).

Connected to this vegetation, the TACF have a specific water dynamics, mainly connected to the high frequency of low density rainfall events and frequency of fog, the last generating and additional water input to these ecosystems (Tobón et al. 2008; Tobón and Arroyave 2007; Bruijnzeel 2001; González 2000), through the fog water interception by vegetation and falling down into the forest floor (Tobón et al. 2010b, Richardson et al. 2000) and the fog control on plant transpiration, as the result of decrease on solar radiation and increases on air humidity during fog events (Ferwerda et al. 2000). Additionally, climate conditions of TACF are mainly controlled by altitude and winds from the Pacific and the Atlantic, maintaining constant humidity on both external slopes of the Cordilleras, while conditions are more variable on the inner flanks. Condensation is significant on the upper portions of the inner flanks, and the middle and lower portions of the valleys have a marked bimodal dry–wet pattern resulting from the rain shadow effect (Herzog et al. 2011).

4.3.2 Key Variables that Control Ecohydrological Processes of TACF

Not different from other terrestrial ecosystems, the TACF receive different amounts of precipitation (Espinoza et al. 2009; Emck et al. 2006; DeAngelis et al. 2004; Liebmann et al. 2004; Vuille et al. 2000; Póveda and Mesa 1997; Grubb and Whitmore 1996), ranging from 500 to 5000 mm y^{-1} (Schawe et al. 2008; Sarmiento 2001), depending on their exposure and position within the Andean mountains (Martínez et al. 2011). A long series of data (1998–2004) indicates that precipitation on the eastern side of the Andes ranges from 2060 mm y^{-1} at an altitude of 1960 m to 4400 mm y^{-1} at 3200 m (Oesker et al. 2008), indicating that variability depends on the altitudinal position of each site. A different tendency is found in Colombia, when studying sites located at similar altitude, but exposed to different provenance of air masses. Those TACF exposed to the Pacific air masses, but located far from the ocean (see Table 4.1, site 3), receive intermediate values of precipitation, as compared to those directly exposed to the Amazonia (Table 4.1, site 2), while ecosystems exposed to the plateau valleys receive the least precipitation (Table 4.1, site 1). This implies that when considering data from different sites at similar altitudes, rainfall variability is related to differences in air masses ascending the Andean mountains: some very humid such as those facing the Pacific basin (Póveda et al. 2005; González 2000), and those from the Amazon basin (Espinoza et al. 2009; Tobón 1999).

In this study it was found that rainfall in TACF is characterized by low magnitude and intensity (Tobón 2009; Ataroff 2005; Ataroff 2002), values ranging from to 1575 to 4588 mm y^{-1}, with an average of 2337 mm y^{-1}. Although there is not a clear tendency, rainfall seems to increase with altitude, with some differences between those TACF facing the east or the west, the last receiving more precipitation; contrary to that found by some authors (Espinoza et al. 2009; Bendix et al. 2006; Lara et al. 2003), who indicated that Andean cloud forest facing the east receive greater precipitation than those oriented to the west. Moreover, yearly amounts of precipitation decrease with altitude towards the interior valleys, which is, those TACF exposed to the inter-Andean valleys are relatively dry. This was also found in other cloud forests (DeAngelis et al. 2004; Liebmann et al. 2004; Lara et al. 2003; Vuille et al. 2000). In brief, depending on their position in the altitudinal gradient (Rollenbeck et al. 2008; Harden 2006; Rollenbeck 2006; Bendix et al. 2004; Lauer 1981), and degree of exposure to these cloud masses (Espinoza et al. 2009; Póveda and Mesa 1997), TACF receive certain amounts of vertical and horizontal precipitation, including fog water (Table 4.2).

Fog water seems to contribute significantly to the ecohydrology of TACF ecosystems, directly or indirectly (Tobón et al. 2008; Tobón and Arroyave 2007; Dengel and Rollenbeck 2003; Bruijnzeel and Hamilton 2000; Hamilton et al. 1995; Stadtmüller 1987, 2003). Directly, fog water intercepted by TACF may fall into the forest floor (Bruijnzeel et al. 2010), therefore contributes to net water inputs to the ecosystem. This interception of fog water by TACF is not only an action by forest canopy (Holder 2004), but also by epiphytes, mainly mosses and lichens attached to tree branches and trunks are involved in this process, since they act like a sponge capable of retaining moisture from fog (Tobón et al. 2010a; Köhler et al. 2007; Walker and Ataroff 2005; Mulligan and Jarvis 2000; Cavelier 1996; Veneklaas and Van Ek 1990), to subsequently release it drip to the soil surface (Coxson and Nadkarni 1995; Lovett et al. 1985). Water contribution made by fog also depends on fog density and frequency, wind speed, and the presence of dense vegetation that captures the water present in the fog (Cárdenas, Tobón and Buytaert 2017; Frumau et al. 2011; Villegas et al. 2008; Holder 2004; González 2000; Cavelier 1991).

Average cloud frequency in studied sites in Colombia (2007–2016) was 0.93 ± 0.12 events per day, with the lowest value at site 1 and the highest at site 2 (Table 4.2). This value is higher than that presented by Mulligan et al. (2010), using MODIS cloud climatology, which means that TACF is wrapped at least once a day, being more frequent during the dry periods (January to March and July to September); however those TACF facing a relative dry montane plateau, as the one in Guerrero (site 1), experience much less events, and most of them occur during the wet periods. Here most of the fog events (58%) occurred early in the morning (between 5:00 and 9:00 am), and late in the afternoon and early night (36%).

The Andean altitudinal gradient of precipitation is accompanied by a gradient of fog water inputs, increasing with the altitude, for studied sites (Table 4.2). However values found by Bendix et al. (2008) and Jarvis (2000) in Ecuador and Colombian cloud forests, respectively, are much larger than all values reported by most TACF (see Table 4.2) and elsewhere, in cloud forests (Bruijnzeel et al. 2011). Average value for water inputs through fog deposition to TACF is 295 mm y^{-1} without considering the above-mentioned data. Mostly, large values of fog water inputs can be the result of mixing fog inputs with horizontal precipitation, under events with high wind speed conditions (Frumau et al. 2010), or erroneous potential values predicted from cloud deposition (Jarvis 2000).

Moreover, based on measurements in the three sites in Colombia, the amounts of fog inputs registered through the Juvik method were always higher than those measured with gutters, as net precipitation during fog events (Table 4.2), the last being considering as real fog water inputs to the ecosystems (Bruijnzeel et al. 2011). From the three sites in Colombia, annual inputs of fog water measured with Juvik cylinders averaged 407 mm, while net fog inputs to troughs installed under the forest were only 116 mm y^{-1}, in average (Table 4.2). From field observations these differences mostly relate to total surface area exposed to fog, acting as barriers trapping water from fog, but also the texture of measurements devices and texture of vegetation. Cylinders seem to be more efficient on capturing the very small fog drops, while in the forests, fog mostly gets in contact with treetops, but most of it passes over, thus, very small amounts of fog water can be captured by exposed tree foliage and branches, generating low values of throughfall. Moreover, it was observed in the field, that very little of the fog that comes in, enters the forests, therefore little or almost no water is trapped by undergrowth or leaves and branches below the treetops. This was confirmed by data from the leaf wetness sensors, when analyzing results only from fog events. Measurements show that those installed on top of the three, got wet during most fog events, but only during prolonged and dense fog events, those installed 2 m and 5 m below the tree crown, get wet (Fig. 4.2).

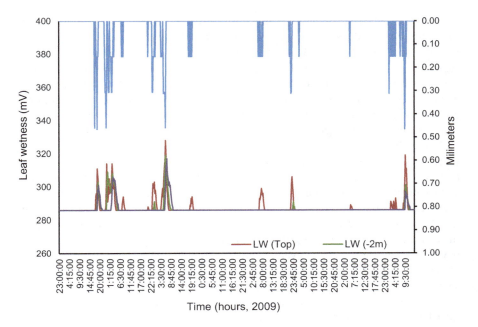

Fig. 4.2 Leaf wetness through forest canopy during fog events in Chingaza, TACF

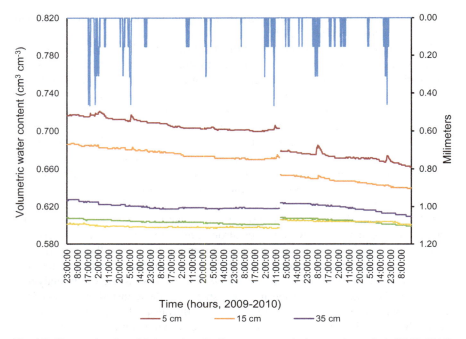

Fig. 4.3 Temporal and spatial dynamics of soil water content during two dry periods (2009–2010), but with water inputs through fog interception (Chingaza, TACF)

Moreover when analyzing data on soil moisture it is observed that during very long duration and dense fog events, soil moisture slightly increased at soil surface (5 and 35 cm depth) (Fig. 4.3); however during most fog events, although data was registered as leaf wetness, and fog water inputs, soil moisture sensors did not show any increase in soil humidity. This is mostly related to the fact that soils in studied TACF in Colombia (Table 4.1) remained either close to field capacity or saturated, almost during the entire studied period, thus small water inputs into the soil surface could not be registered, given the sensibility of TDR sensors (Campell Sci).

Indirectly, the presence of fog covering the TACF intercepts part of the solar radiation, thus reduces direct radiation into the forest, which in turn reduces temperature and increases air humidity (Tobón 2009). These combinations of specific environmental conditions during the fog events reduce plant transpiration and water evaporation from the trees crown (Jarvis and Mulligan 2010). Studies carried out in TACF in Colombia indicated that evapotranspiration during foggy days was reduced in average, 63%, as compared to the value found for open sky day, which is 1.47 mm/day in average for measured events. Noteworthy that, for the same days with fog, inputs by fog drip into the gutters were, in average, of the order of 1.75 mm. It should be stated that during short fog events (shorter than 30 min) or under light fog events (those events were objects can be seen and recognized at a distance shorter than 10 m), evapotranspiration was slightly reduced (lower than 10% as compared

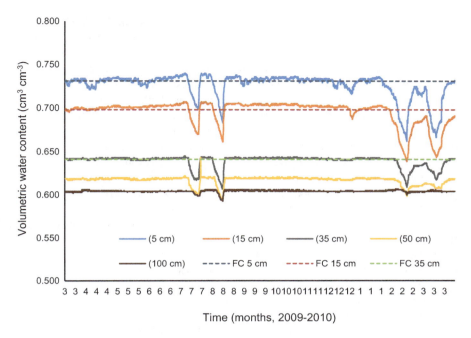

Fig. 4.4 Spatial and temporal dynamics of volumetric water content in soils from Chingaza TACF. Values of field capacity (FC) are also shown for the three first soil layers

to that at open sky) and no inputs of water were registered on the troughs installed under the TACF (unpublished data).

Although the reduction on plant transpiration during foggy events was not directly measured in studied cloud forests (e.g. through sapflow measurements), available data on volumetric soil water content shows that during fog events, soil moisture within the fine root zone remains constant, which implies that there were no plant transpiration during such foggy periods. This could be proved in a foggy day during the dry period (e.g. February 13, 2010) when soil moisture remained constant at all soil depths (Fig. 4.4), while reduction on volumetric water content was observed in those open sky days, values been in the order of 2.89 ± 0.67 mm/day, in the first 50 cm of the soil profile.

Rainfall interception by forests is defined as the amount of rain that is intercepted by vegetation and drains into the soil surface or evaporates (Fleischbein et al. 2005; Tobón and Arroyave 2008; Tobón et al. 2000a; Veneklaas and Ek 1990; Vis 1986). The fraction reaching the soil surface is called net precipitation and the fraction that evaporates becomes the net loss of water from forest ecosystems together with plant transpiration. Moreover, under forests, the litter lying on the forest floor intercepts part of the water entering forest ecosystems, as forest floor interception (Tobón et al. 2000b). These have been considered as the most important ecohydrological processes in forest ecosystems (Savenije 2004; Gerrits et al. 2007).

Reported values of rainfall intercepted by TACF range from 7 to 52% of gross rainfall and fog inputs measured with devices outside the forests (Table 4.2), with an average value of 28%. Large values of rainfall interception reported for TACF are rare and mostly related to the specific methodologies used to measured net precipitation and the number of rain gauges located under the forest. When low number of gauges are used to measure net precipitation (Fonseca and Ataroff 2005; Ataroff 2002; Ataroff and Rada 2000; Wolf 1993), the amount of water intercepted by the forest can be either, very low or very high, depending on rainfall amounts and characteristics, and canopy characteristics as the fraction of crown cover or leaf area index (Tobón et al. 2000a). Values can also be dependent on fog frequency and drop density in fog, forest disturbance, and abundance of epiphytes (Hofstede et al. 1993; Bruijnzeel 2001; Köhler et al. 2007, 2011; Tobón et al. 2010a). For those cloud forests were fog frequency and density is high, amounts of rainfall interception seem to be low (Mulligan et al. 2010; Hölscher et al. 2004). Contrary, the presence of large amounts of epiphytes biomass in some cloud forests seems to be capable of interception large amounts of rainfall (Oesker et al. 2010; Tobón et al. 2010a; Tobón and Arroyave 2007; Hölscher et al. 2004) although the real ecohydrological contribution of epiphytes in montane tropical forests is still under discussion (Veneklaas and Van Ek 1990; Coxson and Nadkarni 1995; Clark et al. 2014; Hafkenscheid 2000; Köhler et al. 2007, 2011), given their capacity to capture water from fog (Tobón et al. 2010a).

Unpublished data from Colombian sites (Table 4.1) has shown that, when using same methodology (e.g. same devices and same number of devices, with relocation), amounts of rainfall interception (including fog) are much lower (13% of gross precipitation plus fog inputs) than reported values for all TACF (Table 4.2) and depend on leaf area index and fog incidence. As indicated above, during fog events, the canopy gets wet (Fig. 4.2), therefore less water from rain is required to saturate the forest canopy in these TACF. Consequently, low values reported for these sites (Table 4.2) will clearly change if the fog was not present.

Soils from TACF have received little attention, including poor soil classification and when those exists, they are general soil studies, to the point that most studies refer to them with the general words "tropical soils" (IGAC 2004). This becomes worse for studies on the hydro-physical properties and soil water dynamics (Tobón et al. 2010b), as compared with other terrestrial ecosystems, including tropical montane ecosystems (Tobón et al. 2010b; Tobón 2009; Podwojewski et al. 2002). This is surprising considering the importance of TACF as water supplier for a large Andean population and the potential threat posed to this function by land use changes and climate change (Eller et al. 2015; Beninston 2003; Walther et al. 2002). In spite of this fact, the few studies carried out in TACF that included soil description, coincide that these have high organic matter content, which is connected to vegetation, including the abundance of epiphytes (Köhler et al. 2007, 2011; Nadkarni 1984), the low temperatures, high soil moisture values throughout the year (Tobón 2009), and the stable organo-metallic complexes that reduce the biological activity and therefore decrease the rate of organic matter mineralization (Bohlman et al. 1995; Vance and Nadkarni 1990).

Soil water content not only determines the excess of water that becomes discharge or groundwater recharge, but also the water availability for plants and transfer processes between soil–plant–atmosphere (Famiglietti et al. 1998), however temporal and spatial variability may constraint a proper role of soil moisture on those processes (Dobriyal et al. 2012). Moreover, soil moisture varies considerably with depth, mainly on those heterogeneous soils with contrasting textures between successive soil layers (Hincapie and Germann 2009; Hincapié and Tobón 2012; Neary et al. 2009; Lazarev et al. 2005), while soil moisture shows more stability on high organic matter soils.

Data from studies listed in Table 4.2 indicate that average soil water content varies considerably between sites. Average value for those sites where soil moisture was measured is around 45%, implying that almost half of the soil components is water. However this percentage may vary depending on the rainfall temporal conditions. During the dry seasons, the lowest values observed were around 23%, while during the wet seasons, soil remains near field capacity or saturated, depending on the frequency of rainfall/fog events.

From the studies carried out by this author in the three sites in Colombia, soil moisture showed to be persistently high, remaining close to field capacity throughout the year, except for those short dry periods when soil water content gently decreases below field capacity (Fig. 4.3). Contrary, during the very wet periods (March to June and October to December), studied TACF soils remain between field capacity and saturation. This unique behavior seems to be related, first to the continuous low intensity rainfall and the frequency of fog events, and second, to the large amounts of organic matter content in these soils, the presence of volcanic ashes or both. At Guerrero and Chingaza cloud forests, soils present large amounts of soil organic matter (between 17.8 and 25%), which partly controlled their soil moisture dynamics, while in Frontino cloud forests, the water dynamics are controlled by both, soil organic matter (26.5%) and volcanic ashes (unpublished data).

Contrary to that indicated by Jung et al. (2010) on limited moisture environments, the high soil water content in TACF indicates that plant transpiration is not limited by soil moisture, at least not for water deficit. However, no studies carried out in TACF have shown the effects of water stress under saturation condition. A study from different forest types showed that water excess in the soil root zone limited plant transpiration (Purdy et al. 2018), therefore affecting plant growing and physiological forest dynamics (Fig. 4.3).

In studied TACF in Colombia, changes in soil water content were larger for the first 0.4 m, but only during the short dry periods, when soil water content slightly decreased from field capacity, but never reached the wilting point, at any soil depth (Fig. 4.4). This was in line with fine root distribution, whose results indicated that most fine roots in studied TACF concentrate in the topsoil. Smallest variability was observed during wet periods, when soil moisture remained close to field capacity, and to some extent, close to saturation, mostly controlled by rainfall and possibly decreasing evapotranspiration due to the presence of low clouds and fog, less radiation, high air humidity, and permanent low temperatures. Additionally, as indicated above, during some fog events, soil moisture slightly increased, which was clear during the short dry periods (Fig. 4.3); however it was observed through the through-fall measurements that fog water inputs to TACF occur throughout the year, but increases on soil water content cannot be visible, as soil moisture remains very high, close, either to field capacity or saturation, therefore small water inputs are not reflected on soil water content increases. This phenomenon may occur in most organic or volcanic soils, where moisture remains very close to, either field capacity or saturation, thus, small water inputs through fog deposition cannot be observed by measuring soil moisture.

Finally, data from studied TACF in Colombia shows that although small variations on soil moisture occur with soil depth and through the soil profiles (mainly during the short dry periods), an analysis of the root mean square error (RMSE) of the relative differences in volumetric soil water content measurements at the different depths ad sites showed that there were some temporal stability of soil water content, throughout the soil profile and time. However, soil water content was temporarily more stable at those soil horizons richer on organic matter than on those volcanic layers. This can be explained by the large amounts of soil organic matter, with large water storage capacity (Tobón et al. 2010b; Hincapié and Tobón 2012) and the specific climatic conditions of TACF, notably, the frequency of fog, which, as indicated above, reduces plant transpiration, and by dripping into the forest floor, add some extra water to the soil surface, thus, any extra water inputs may compensate for the outputs by forest transpiration.

Although the low number of studies that included calculations of evapotranspiration (ETa), values in Table 4.2 indicated that most TACF have low rates of evapotranspiration, which are compared with values from most cloud forests (Silva et al. 2017; Rittera and Regalado 2017; Bruijnzeel 2001, 2004), but differ from other forest types (Tanaka et al. 2008; Tobón 1999). ETa in TACF averages 553 mm y^{-1}, with larger values for sites below 2300 masl. Highest values are mostly related to differences on climatic conditions within sites, but also on fog frequency. In studied sites in Colombia, values of ETa were lower for Frontino and Chingaza sites, which mostly relates to fog frequency. The low temperature and moist conditions provided by fog may reduce leaf temperature and vapor pressure deficit thus enhancing stomata conductance (Smith and McClean 1989).

Data on evapotranspiration of TACF ecosystems shows that there are two main tendencies on the magnitude of ET. The first is related to the occurrence of humid and dry periods in the tropics. Larger values are observed during the dry periods or less-humid periods, and on open sky days. However when there were some fog events in these periods, ET was reduced considerably, due to that fog reduced vapor pressure deficit. The second relates to fog presence and frequency. Leaf wetness, connected to either rainfall or fog events (Fig. 4.5), suppress plant transpiration (Gotsch et al. 2014). This implies that in sites where either rainfall events are frequent or fog events are persistent and of long duration, tree transpiration must be reduced by both, the increase in leaf wetness and the changes in meteorological conditions connected to the presence of fog. The last was observed in the studied sites in Colombia, showing that during rainfall events and/or dense fog events lasting for at least 1 h, forest canopy exhibited some leaf wetness, being these larger as larger and dense was the event (Fig. 4.2).

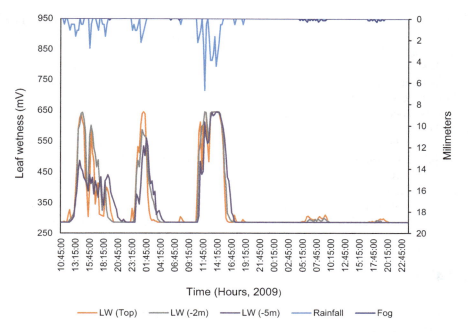

Fig. 4.5 Spatial and temporal dynamics of leaf wetness during rainfall and fog events in Chingaza TACF

Moreover, provenance of air masses and their moisture seems to control evapotranspiration from TACF ecosystems. When air masses come from areas relatively wet (Pacific and Amazon basins), ET is lower (see sites 2 and 3) than when these air provenance is from a relatively dry valley, as the case site 1 (Guerrero). In this case, saturated vapor pressure restricts evapotranspiration, no matter if soil water is available for plant transpiration. This indicates that contrary to the global tendency of land evapotranspiration decline found by (Jung et al. 2010), ETa data from TACF shows that actual evapotranspiration rates are not controlled by soil moisture, which remains near to field capacity, throughout the year, but for atmospheric control parameters as fog, the frequency of rainfall events, and wet air masses, reducing incident solar radiation and temperature, increasing air humidity, and wetting tree crowns (Fig. 4.5).

Water balance for a given ecosystem allows us for a better understanding on the relationships between water inputs and outputs, specifically in cloud forests, where no discharge has been measured, where its calculation renders the water yields. Water cycle within tropical cloud forests is a very dynamic and complex process, both spatially and temporally, therefore accuracy of water balance is highly dependent on precision of measured hydrometeorological variables, as rainfall and fog inputs, rainfall water interception, evapotranspiration, and soil moisture temporal dynamics, preferably on annual or multiannual basis.

Applying the general equation for the water balance in those sites cited in Table 4.2, water yield from TACF ecosystems ranges between 0.34 and 0.73, with an average value of 0,48 ± 0.16, implying that for each millimeter of water entering to these ecosystems, either though rainfall or fog, 0.48 mm of water leaves the ecosystem through discharge. This value is lower than the average value presented by Tobón (2009) for páramo ecosystem, but higher than values found for tropical rain forests (Jones et al. 2017; van Dijk et al. 2012; McJannet et al. 2007) and much higher than those from tropical dry forest (Allen et al. 2017; Caldwell et al. 2016). It should be stated here that in most hydrological studies reported for TACF (Table 4.2), no account has been taken for fog inputs, and in most cases evapotranspiration has been evaluated as potential evapotranspiration, which may generate errors on the final calculation of water yield. In those studies where actual evapotranspiration and fog inputs were considered, the average water yield by TACF is 0.56, which is similar to that value found for tropical cloud forests (Tobón 2009).

4.4 Conclusions

Within the framework of this study it is clear that vegetation of TACF, including the epiphytes, modifies water inputs to the ecosystems: 28% of total water inputs is intercepted and evaporated from the forest canopy, which becomes in a net loss from the ecosystem, while fog water is also intercepted and new inputs of water are added to the forest floor, whose proportion is 560 mm y^{-1}, in average. Moreover, authors indicate that during fog events, transpiration by plants is depressed.

Ecohydrologically, TACF are specifically influenced by four important factors: Specific climate conditions, as low temperatures and high air humidity, relatively high and continuous rainfall inputs, with almost no presence of dry periods, fog presence, in some sites frequently, and consequently low evapotranspiration rates. These variables significantly control the ecohydrological functioning of undisturbed TACF.

Finally, a better understanding of TACF ecohydrology can be reached if future studies include real fog inputs to these ecosystems, preferably measured through net precipitation methods, and soil moisture can be also measured. Given the large spatial and temporal variability, it will be preferred if studies can be made for long periods on different TACF sites.

References

Allen K, Dupuy JM, Gei MG, Hulshof C, Medvigy D, Pizano C, Salgado-Negret B, Smith CM, Trierweiler A, Van Bloem SJ, Waring BG, Xu X, Powers JS (2017) Will seasonally dry tropical forests be sensitive or resistant to future changes in rainfall regimes? Environ Res Lett 12(2):023001

Ataroff M (1998) Importance of cloud water in Venezuelan Andean cloud forest water dynamics. In: Schemenauer RS, Bridgman HA (eds) Proceedings of the first international conference on fog and fog collection. International Development Research Centre, Ottawa, pp 25–28

4 Ecohydrology of Tropical Andean Cloud Forests

Ataroff M (2002) Precipitación e intercepción en ecosistemas boscosos de los andes Venezolanos. Ecotropicos 15(2):195–202. Sociedad Venezolana de Ecología

Ataroff M (2005) Estudios de dinámica hídrica en la Selva Nublada de La Mucuy, Andes de Venezuela. In: Ataroff M, Silva JF (eds) Dinámica Hídrica en Sistemas Neotropicales. ICAE. Univ. Los Andes, Mérida

Ataroff M, Rada F (2000) Deforestation impact on water dynamics in a Venezuelan Andean cloud forest. Ambio 29:440–444

Barradas VL, Cervantes-Pérez J, Ramos-Palacios R, Puchet-Anyul C, Vázquez-Rodriguez P, Granados-Ramírez R (2010) Meso-scale climate change in the central mountain region of Veracruz State, Mexico. In: Bruijnzeel LA, Scatena FN, Hamilton LS (eds) Tropical montane cloud forests. Science for conservation and management. Cambridge University Press, Cambridge, pp 549–556

Beck E, Bendix J, Kottke I, Makeschin F, Mosandl R (2008) Gradients in a tropical mountain ecosystem of Ecuador. Ecological studies (analysis and synthesis), vol 198. Springer, Berlin, 543 p

Bendix J, Fabiany P, Rollenbeck R (2004) Gradients of fog and rain in a tropical montane cloud forest of southern Ecuador and its chemical composition. In: Proceedings 3rd Int. Conf. on Fog. Fog Collection and Dew, 11–15 Oct. 2004. Cape Town

Bendix J, Rollenbeck R, Reudenbach C (2006) Diurnal patterns of rainfall in a tropical Andean valley of southern Ecuador as seen by a vertically pointing K-band Doppler radar. Int J Climatol 26(6):829–846

Bendix J, Rollenbeck R, Richter M, Fabian P, Emck P (2008) Gradual changes along the altitudinal gradients. The climate. In: Beck E, Bendix J, Kottke I, Makeschin F, Mosandl R (eds) Gradients in a tropical mountain ecosystem of Ecuador. Ecological studies, vol 198. Springer, Berlin, pp 63–73

Beninston M (2003) Climatic change in mountain regions: a review of possible impacts. Climate Change 59:5–31

Bohlman SA, Matelson TJ, Nadkarni NM (1995) Moisture and temperature patterns of canopy humus and forest floor soil of a Montane Cloud Forest, Costa Rica. Biotropica 27(1):13–19

Brown JH, Lomolino MV (1998) Biogeography, 2nd edn. Sinauer Associates, Sunderland, MA, xii + 691 pp

Bruijnzeel LA (2001) Hydrology of tropical montane cloud forests: a reassessment. Land Use Water Resour Res 1:1D1–1D18

Bruijnzeel LA (2004) Hydrological functions of tropical forests: not seeing the soil for the trees? Agric Ecosyst Environ 104:185–228

Bruijnzeel LA, Hamilton LS (2000) Decision time for cloud forests, IHP humid tropics programme series, vol 13. Paris, UNESCO Division of Water Sciences. http://sea.unepwcmc.org/forest/cloudforest/index.cfm. Accessed Oct 2019

Bruijnzeel LA, Scatena FN, Hamilton LS (2010) Tropical montane cloud forests. Science for conservation and management. Cambridge University Press, Cambridge, 342 pp

Bruijnzeel LA, Mulligan M, Scatena F (2011) Hydrometeorology of tropical montane cloud forests: emerging patterns. Hydrol Process 25:465–498

Bruinsma J (2003) World agriculture: towards 2015/2030. A FAO perspective. Earthscan, London, 432 pp

Burbano-Garcés ML, Figueroa-Casas A, Peña M (2014) Bulk precipitation, throughfall and stemflow deposition of N-NH4+, N-NH3 and N-NO3 - in an Andean forest. J Trop For Sci 26(4):446–457

Caldwell PV, Minia FC, Elliot KJ, Swank WT, Brantley ST, Laseter SH (2016) Declining water yield from forested mountain watersheds in response to climate change and forest mesophication. Glob Chang Biol 22:2997–3012

Cárdenas MF, Tobón C, Buytaert W (2017) Contribution of occult precipitation to the water balance of páramo ecosystems in the Colombian Andes. Hydrol Process 31:4440–4449

Cavalier J, Goldstein G (1989) Mist and fog interception in Elfin cloud forests in Colombia and Venezuela. J Trop Ecol 5:309–322

Cavelier J (1996) Environmental factors and ecophysiological processes along altitudinal gradients in wet tropical mountains. In: Mulkey SS, Chazdon RL, Smith AP (eds) Tropical forest plant ecophysiology. Chapman and Hall, New York, pp 399–439

Clark K, Torres MA, West A, Hilton R, New M, Horwath A, Fisher J, Rapp J, Robles A, Caceres A, Malhi Y (2014) The hydrological regime of a forested tropical Andean catchment. Hydrological. Earth Syst Sci 18:5377–5397

Condit R, Aguilar S, Hernandez A, Perez R, Lao S, Pyke C (2005) Spatial changes in tree composition of high-diversity forests: how much is predictable? In: Bermingham E, Dick CW, Moritz C (eds) Tropical forests: past, present and future. University of Chicago Press, Chicago, pp 271–294

Coxson DS, Nadkarni NM (1995) Ecological role of epiphytes in nutrient cycles. In: Lowma MD, Nadkarni NM (eds) Forest canopies. Academic Press, Nueva York

Cuesta F, Peralvo M, Valerazo N (2009) Los bosques montanos de los Andes Tropicales. Una evaluación regional de su estado de conservación y de su vulnerabilidad a efectos del cambio climático. Programa Regional ECOBONA-INTERCOOPERATION, 74 pp. www.bosquesandinos.info

Cavelier J, Solis D, & Jaramillo M (1996) Fog interception in montane forests across the central cordillera of panama. Journal of Tropical Ecology 12(3):357–369. Retrieved September 3, 2020, from http://www.jstor.org/stable/2560056

CAR - Corporación Autónoma Regional de Cundinamarca (2001). Atlas Ambiental. http://hdl.handle.net/20.500.11786/36139

DeAngelis CF, McGregor GR, Kidd C (2004) A 3-year climatology of rainfall characteristics over tropical and subtropical South America based on tropical rainfall measuring mission precipitation radar data. Int J Climatol 24:385–399

Dengel S, Rollenbeck R (2003) Methods of fog quantification in a tropical mountain forest of southern Ecuador. Fog Newsletter, 15

Dobriyal P, Qureshi A, Badola R, Hussain SA (2012) A review of the methods available for estimating soil moisture and its implications for water resource management. J Hydrol 4:110–117

Dietz J, Hölscher D, Leuschner C, Hendrayanto H (2006) Rainfall partitioning in relation to forest structure in differently managed montane forest stands in Central Sulawesi, Indonesia. Forest Ecology and Management 237:170–178. https://doi.org/10.1016/j.foreco.2006.09.044

Eller CE, Burgess SSO, Oliveira RS (2015) Environmental controls in the water use patterns of a tropical cloud forest tree species: Drimys brasiliensis (Winteraceae). Tree Physiol 35:387–399

Emck P, Moreira-Muñoz A, Richter M (2006) El clima y sus efectos en la vegetación. In: Moraes M et al (eds) Botánica Económica de los Andes Centrales. Universidad Mayor de San Andrés, La Paz, pp 11–36

Espinoza JC, Ronchail J, Guyot JL, Filizola N, Noriega L, Ordoñez JJ, Pombosa R, Romero H (2009) Spatio – temporal rainfall variability in the Amazon Basin countries (Brazil, Peru, Bolivia, Colombia and Ecuador). Int J Climatol 29:1574–1594

Famiglietti JS, Rudnicki JW, Rodell M (1998) Variability in surface soil moisture content along a hillslope transect: Rattlesnake Hill, Texas. J Hydrol 210:259–281

Ferwerda W, Hadeed L, McShane T, Rietbergen S, Stolton S, Dudley D (2000) Bosques Nublados Tropicales Montanos. WWF International/IUCN. The World Conservation Union, 63 p

Fittkau E, Illies J, Klinge H, Schwab G, Sioli H (1968) Biogeography and ecology in South America. Junk Publishers, The Hague

Fleischbein K, Wilcke W, Goller R, Boy J, Valarezo C, Zech W, Knoblich K (2005) Rainfall interception in a lower montane forest in Ecuador: effects of canopy properties. Hydrol Process 19:1355–1371

Fleischbein K, Wilcke W, Valarezo C, Zech W, Knoblich K (2006) Water budgets of three small catchments under montane forest in Ecuador: experimental and modelling approach. Hydrol Process 20:2491–2507

Fonseca H, Ataroff M (2005) Dinámica Hídrica en la selva nublada de la cuenca alta del Rio Cusiana y un pastizal de reemplazo, Cordillera Oriental, Colombia. In: Ataroff M, Silva JF (eds) Dinámica Hídrica en Sistemas Neotropicales. ICAE, Univ. Los Andes, Mérida

Fontúrbel F (2002) Los bosques andinos: reseña biogeográfica y elementos representativos. Long Rev. 10:12–19. biologia.org

Föster P (2001) The potential negative impacts of global climate change on tropical montane cloud forests. Earth Sci Rev 55:73–106

Frumau A, Schmid S, Burkard R, Bruijnzeel LA, Tobón C, Calvo J (2010) Fog gauge performance as a function of wind speed in northern Costa Rica. In: Bruijnzeel LA, Juvik J, Scatena FN, Hamilton LS, Bubb P (eds) Forests in the mist: science for conservation and management of tropical montane cloud forests. Cambridge University Press, Cambridge, pp 294–301

Frumau KFA, Burkard R, Schmid S, Bruijnzeel LA, Tobón, C, and Calvo-Alvarado, J (2011) A comparison of the performance of three types of passive fog gauges under conditions of wind-driven fog and precipitation. Hydrological Processes 25(3):374–383

Garcia CF (2007) Regulación hídrica bajo tres coberturas vegetales en la cuenca del rio San Cristóbal, Bogotá D.C. Rev Colomb For 10(20):127–147

Garreaud RD (2009) The Andes climate and weather. Adv Geosci 7(1–9):2009

Gerrits AMJ, Savenije G, Hoffmann L, Pfister L (2007) New technique to measure forest floor interception – an application in a beech forest in Luxembourg. Hydrol Earth Syst Sci 11:695–701

Gómez-Peralta D, Oberbauer SF, McClain ME, Philippi TE (2008) Rainfall and cloud-water interception in tropical montane forests in the eastern Andes of Central Peru. For Ecol Manag 255:1315–1325

González J (2000) Monitoring cloud interception in a tropical montane forest of the southwestern Colombian Andes. Adv Environ Monit Model 1:97–117

Gotsch SG, Asbjornsen H, Holwerda F, Goldsmith GR, Weintraub AE, Dawson TE (2014) Foggy days and dry nights determine crown-level water balance in a seasonal tropical montane cloud forest. Plant Cell Environ 37:261–272

Grubb PJ, Whitmore TC (1996) A comparison of montane and lowland rain forest in ecuador II: the climate and its effects on the distribution and physiognomy of the forests. J Ecol 54(2):303–333

Gentry AH (1991) The distribution and evolution of climbing plants. In: The Biology of Vines (eds F.E. Putz & H.A. Mooney). Cambridge University Press, Cambridge, pp. 3– 49.

Hafkenscheid RRLJ (2000) Hydrology and biogeochemistry of tropical montane rain forests of contrasting stature in the Blue Mountains, Jamaica. PhD Dissertation, Vrije Universiteit Amsterdam, Ámsterdam

Hamilton LS, Juvik JO, Scatena FN (1995) The Puerto Rico tropical cloud forest symposium: introduction and workshop synthesis. In: Hamilton LS, Juvik JO, Scatena FN (eds) Tropical montane cloud forests. Springer-Verlag, New York, pp 1–23

Harden C (2006) Human impacts on headwater fluvial systems in the northern and Central Andes. Geomorphology 79:249–263

Herzog SK, Martínez R, Jørgensen PM, Tiessen H (2011) Climate change and biodiversity in the Tropical Andes. Inter-American Institute for Global Change Research (IAI) and Scientific Committee on Problems of the Environment (SCOPE), Brazil

Hincapie I, Germann PF (2009) Impact of initial and boundary conditions on preferential flow. J Contam Hydrol 104:67–73

Hincapié E, Tobón C (2012) Dinámica del Agua en Andisoles Bajo Condiciones de Ladera. Revista de la Facultad Agronómica de Medellín, Universidad Nacional de Colombia. Review of the Facilative National. Agric Medellín 65(2):6771–6783

Hofstede RGM, Wolf JH, Benzing DH (1993) Epiphyte biomass and nutrient status of a Colombian upper montane rain forest. Selbyana 14:37–45

Holder CD (2004) Rainfall interception and fog precipitation in a tropical montane cloud forest of Guatemala. For Ecol Manag 190:373–384

Hölscher D, Köhler L, Van Dijk AIJM, Bruijnzeel LA (2004) The importance of epiphytes to total rainfall interception by a tropical montane rain forest in Costa Rica. J Hydrol 292:308–322

Holwerda F, Bruijnzeel LA, Muñoz-Villers LE, Equihua M, Asbjornsen H (2010) Rainfall and cloud water interception in mature and secondary lower montane cloud forests of Central Veracruz, Mexico. J Hydrol 384:84–96

Hursch CR, Brater EF (1941) Separating storm-hydrographs from small drainage-areas into surface- and subsurface-flow. Trans Am Geophys Union Part 3(22):863–871

Ives JD, Messerli B, Spiess E (1997) Mountains of the world: a global priority. In: Messerli B, Ives JD (eds) Mountains of the world: a global priority. Estados Unidos y Carnforth, Reino Unido, Parthenon Publishing Group, Nueva York, pp 1–15

IGAC – Instituto Geográfico Agustín Codazzi (2004) Levantamiento de suelos y zonificación de tierras. Estudio de suelos regionales. Bogotá, D.E. 537 p. http://www2.igac.gov.co/igac_web/contenidos/plantilla_general_titulo_contenido.jsp?idMenu=129

Jackson RB, Carpenter SR, Dahm CN, McKnight DM, Naiman RJ, Postel SL, Running SW (2001) Water in a changing world. Ecol Appl 11:1027–1045

Jarvis A (2000) Quantifying the hydrological role of cloud deposition onto epiphytes in a tropical montane cloud forest, Colombia. http://www.ambiotek.com/herb/hydjar.pdf

Jarvis A, Mulligan M (2010) The climate of tropical montane cloud forests. Hydrol Process 25(3):327–343. https://doi.org/10.1002/hyp.7847

Jones J, Almeida A, Cisneros F, Ioumé A, Jobbágy E, Lara A, Lima WDP, Little C, Llerena C, Silveira L, Villegas C (2017) Forests and water in South America. Hydrol Process 31:972–980

Jung M, Reichstein M, Ciais P, Seneviratne SI, Sheffield J, Goulden ML, Bonan G, Cescatti A, Chen J, de Jeu R, Dolman AJ, Eugster W, Gerten D, Gianelle D, Gobron N, Heinke J, Kimball J, Law BE, Montagnani L, Mu Q, Mueller B, Oleson K, Papale D, Richardson AD, Roupsard O, Running S, Tomelleri E, Viovy N, Weber U, Williams C, Wood E, Zaehle S, Zhang K (2010) Recent decline in the global land evapotranspiration trend due to limited moisture supply. Nature 467:951–954

Juvik JO, Ekern PC (1978) A climatology of mountain fog on Mauna Loa, Hawaiï Island, technical report, vol 118. Water Resources Research Center, University of Hawaiï, Honolulu

Kaser G, Georges C, Juen I, Moelg T (2005) Low latitude glaciers: unique global climate indicators and essential contributors to regional fresh water supply. A conceptual approach. Global change and mountain regions: a state of knowledge overview, advances in global change research, vol 23, Huber U, Bugmann HKM, Reasoner MA. Kluwer Publishers: New York; 185–196

Köhler L, Tobón C, Frumau A, Bruijnzeel L (2007) Biomass and water storage dynamics of epiphytes in old-growth and secondary montane cloud forest stands in Costa Rica. Plant Ecol 193(2):171–184

Köhler L, Hölscher D, Bruijnzeel LA, Leuschner C (2011) Epiphyte biomass in Costa Rican old-growth and secondary montane rain forests and its hydrological significance. In: Tropical montane cloud forests: science for conservation and management. Cambridge University Press, Cambridge, pp 268–274. https://doi.org/10.1017/CBO9780511778384.029

Lara M, Rollenbeck R, Fabian P, Bendix J (2003) Relaciones entre precipitación y vegetación en el bosque tropical de montaña. 2da conferencia. Ecología de bosques tropicales. Loja, Ecuador, 2002

Lauer W (1981) Ecoclimatological conditions of the Páramo belt in the tropical high mountains. Mt Res Dev 1:209–221

Lawton RO, Nair US, Pielke RA Sr, Welch RM (2001) Climatic impact of tropical lowland deforestation on nearby montane cloud forests. Science 294:584–587

Lazarev YN, Petrov PV, Tartakovsky D (2005) Interface dynamics in randomly heterogeneous porous media. Adv Water Resour 28:393–403

León Peláez JD, González Hernández MI, Gallardo Lancho JF (2010) Distribución del agua lluvia en tres bosques altoandinos de la cordillera Central de Antioquia, Colombia. Rev Fac Natl Agron 63:5319–5336

Liebmann B, Kliadis G, Vera C, Saulo A, Carvalho L (2004) Subseasonal variations of rainfall in South America in the vicinity of the low-level jet east of the Andes and comparison to those in the South Atlantic Convergence Zone. J Climatol 17:3829–3842

Lovett GM, Lindberg SE, Richter DD, Johnson DW (1985) The effects of acidic deposition on cation leaching from three deciduous forest canopies. Can J For Res 15:1055–1060

Martínez R, Ruiz D, Andrade M, Blacutt L, Pabón D, Jaimes E, León G, Villacís M, Quintana J, Montealegre E, Euscátegui C (2011) Synthesis of the climate of the Tropical Andes. In: Herzog S, Martinez R, Jørgensen P, Tiessen H (eds) Climate change and biodiversity in the tropical

4 Ecohydrology of Tropical Andean Cloud Forests

Andes. Inter-American Institute for Global Change Research (IAI) and Scientific Committee on Problems of the Environment (SCOPE). https://doi.org/10.13140/2.1.3718.4969

McJannet D, Wallace J, Fitch P, Disher M, Reddell P (2007) Water balance of tropical rainforest canopies in north Queensland. Aust Hydrol Process 25:3473–3484

Moser G, Röderstein M, Soethe N, Hertel D, Leuschner C (2008) Altitudinal changes in stand structure and biomass allocation of tropical mountain forests in relation to microclimate and soil chemistry. In: Beck E et al (eds) Gradients in a tropical mountain ecosystem of Ecuador. Ecological studies, vol 198. Springer, Berlin, pp 229–242

Mulligan M, Jarvis A (2000) Laboratory simulation of cloud interception by epiphytes and implication for hydrology of the Tambito experimental cloud Forest, Colombia. J Hydrol 403:853–858

Mulligan M, Jarvis A, González J, Bruijnzeel LA (2010) Using 'biosensors' to elucidate rates and mechanisms of cloud water interception by epiphytes, leaves, and branches in a sheltered Colombian cloud forest. In: Bruijnzeel LA, Scatena FN, Hamilton LS (eds) Tropical montane cloud forests. Science for conservation and management. Cambridge University Press, Cambridge, pp 249–260

Muñoz-Villers LE, Holwerda F, Gómez-Cárdenas M, Equihua M, Asbjornsen H, Bruijnzeel LA, Marín-Castro BE, Tobón C (2012) Water balances of old-growth and regenerating montane cloud forests in Central Veracruz, Mexico. J Hydrol 462–463:53–66. https://doi.org/10.1016/j.jhydrol.2011.01.062

Myers N, Mittermeier R, Mittermeier C, da Fonseca G, Kent J (2000) Biodiversity hotspots for conservation priorities. Nature 403:853–858

Nadkarni NM (1984) Epiphyte biomass and nutrient capital of a Neotropical Elfin Forest. Biotropica 16:249–256

Nadkarni NM, Wheelwright NT (2000) Ecology and natural history of a tropical montane cloud forest, Monteverde, Costa Rica. Oxford University Press, New York

Neary DG, Ice GG, Jackson CR (2009) Linkages between forest soils and water quality and quantity. For Ecol Manag 258:2269–2281

Oesker M, Homeier J, Dalitz H (2008) Spatial heterogeneity of throughfall quantity and quality in tropical montane forests in southern Ecuador. Second international symposium mountains in the mist: science for conserving and managing tropical montane cloud forest. Hawaii Preparatory Academy (HPA), Waimea (July 27–August 2, 2004)

Oesker M, Homeier J, Dalitz H, Bruijnzeel LA (2010) Spatial heterogeneity of throughfall quantity and quality in tropical montane forests in southern Ecuador. In: Bruijnzeel LA, Scatena FN, Hamilton LS (eds) Tropical montane cloud forests. Science for conservation and management. Cambridge University Press, Cambridge, pp 393–401

Pielou EC (1979) Biogeography. Wiley, Hoboken, 351 p

Podwojewski P, Poulenard J, Zambrana T, Hofstede R (2002) Overgrazing effects on vegetation cover and properties of volcanic ash soil in the páramo of Llangahua and La Esperanza (Tungurahua, Ecuador). Soil Use Manag 18:45–55

Pounds JA, Fogden MPL, Campbell JH (1999) Biological response to climate change on a tropical mountain. Nature 398:611–615

Póveda G, Mesa OJ (1997) Feedbacks between hydrological processes in tropical South America and large-scale ocean–atmospheric phenomena. J Clim 10:2690–2702

Póveda G, Mesa OJ, Salazar LF, Arias PA, Moreno HA, Vieira SC, Agudelo PA, Toro VG, Álvarez JF (2005) The diurnal cycle of precipitation in the tropical Andes of Columbia. Mon Weather Rev 133:228–240

Purdy A, Fisher JB, Goulden ML, Colliander A, Halverson G, Tuc K, Famigliettia J (2018) SMAP soil moisture improves global evapotranspiration. Remote Sens Environ 249:1–14

Rada F (2002) Los bosques andinos: reseña biogeográfica y elementos representativos. Biologia 10:1–16

Ramírez B, Teuling A, Ganzeveld L, Hegger Z, Leemans R (2017) Tropical montane cloud forests: hydrometeorological variability in three neighbouring catchments with different forest cover. J Hydrol 552:151–167

Richardson BA, Richardson MJ, Scatena FN, McDowell WH (2000) Effects of nutrient availability and other elevational changes on bromeliad populations and their invertebrate communities in a humid tropical forest in Puerto Rico. J Trop Ecol 16:167–188

Rittera A, Regalado C (2017) Tree stomata conductance estimates of a wax myrtle-tree heath (fayal-brezal) cloud forest as affected by fog. Agric For Meteorol 247:116–130

Rodríguez-Iturbe I (2000) Ecohydrology: a hydrologic perspective of climate-soil-vegetation dynamics. Water Resour Res 36:3–9

Rollenbeck R (2006) Variability of precipitation in the Reserva Biólogica San Francisco, Southern Ecuador. Lyonia 9:43–51

Rollenbeck R, Bendix J, Fabian P (2008) Spatial and temporal dynamics of atmospheric water and nutrient inputs in tropical mountain forests of Southern Ecuador. Second international symposium mountains in the mist: science for conserving and managing tropical montane cloud forest. Hawaii Preparatory Academy (HPA), Waimea. July 27–August 2, 2004

Sarmiento FO (2001) Ecuador. In: Kapelle M, Brown AD (eds) Bosques Nublados del Neotrópico. Instituto Nacional de Biodiversidad, INBio, Heredia, 698 p

Savenije HHG (2004) The importance of interception and why we should delete the term evapotranspiration from our vocabulary. Hydrol Process 18:1507–1511

Schawe M, Gerold G, Bach K, Gradstein SR (2008) Hydrometeorologic patterns in relation to montane forest types along an elevational gradient in the Yungas, Bolivia. Second international symposium mountains in the mist: science for conserving and managing tropical montane cloud forest. Hawaii Preparatory Academy (HPA). Waimea (July 27–August 2, 2004)

Silva B, Álava-Núñez P, Strobl S, Beck E, Bendix J (2017) Area-wide evapotranspiration monitoring at the crown level of a tropical mountain rain forest. Remote Sens Environ 194:219–229

Sklenár P, Ramsay PM (2001) Blackwell Science, Ltd diversity of zonal páramo plant communities in Ecuador. Divers Distrib 7:113–124

Smith JM, Cleef A (1988) Composition and origins of the world's tropicalpine floras. J Biogeogr 15:631–645

Smith WK, McClean TM (1989) Adaptive relationship between leaf water repellency, stomatal distribution, and gas exchange. Am J Bot 76:465–469

Stadtmüller T (1987) Cloud forests in the humid tropics. A bibliographic review. United Nations University, Tokyo y CATIE, Turrialba

Stadtmüller T (2003) Forests in watershed management as a means to reduce flood risks–the example of the PROMIC project. Presentation at the conference on "Multifunctional forestry and sustainable water management in development cooperation", 26 February 2003. Berna

Steinhardt U (1979) Studies on the water and nutrient balance of an Andean cloud forest in Venezuela. Gött Bodenkd Ber 56:1–185

Soil Survey Staff (2004) Official soil series descriptions. Available at http://soils.usda.gov/technical/classification/osd/index.html (accessed 10 Feb. 2004, verified 17 Dec. 2007). USDA-NRCS.

Tanaka N, Kume T, Yoshifuji N, Tanaka K, Takizawa H, Shiraki K, Tantasirin C, Tangtham N, Suzuki M (2008) A review of evapotranspiration estimates from tropical forests in Thailand and adjacent regions. Agric For Ecol 148:807–819

Tobón C (1999) Monitoring and modelling hydrological fluxes in support of nutrient cycling studies in Amazonian rain forest ecosystems. Tropenbos series 17, Wageningen, the Netherlands, 169 pp

Tobón C (2009) Los bosques andinos y el agua. Serie Investigación y Sistematización 4. Programa Regional ECOBONA—INTERCOOPERATION— CONDESAN: Quito, Ecuador, 122 pp

Tobón C, Arroyave F (2007) Hidrología de los bosques alto andinos. In: León Peláez JD (ed) Ecología de Bosques Andinos, Universidad Nacional de Colombia. Universidad Nacional de Colombia, Bogota, 213 p

Tobón C, Arroyave F (2008) Hidrología de los bosques Alto Andinos. In: León Peláez JD (ed) Ecología de Bosques Andinos. Universidad Nacional de Colombia, Bogota, 260 p

Tobón M, Bouten C, Sevink W (2000a) Gross rainfall and its partitioning into throughfall, stemflow and evaporative loss in four forest ecosystems in western Amazonia. J Hydrol 237:40–57

Tobón MC, Bouten C, Dekker S (2000b) Forest floor water dynamics and root water uptake in four forest ecosystems in northwest Amazonian. J Hydrol 237:169–183

4 Ecohydrology of Tropical Andean Cloud Forests

Tobón C, Gil G, Villegas C (2008) Aportes de la niebla al balance hídrico de los bosques altoandinos. In: León Peláez JD (ed) Ecología de Bosques Andinos. Universidad Nacional de Colombia, Bogota, 213 p

Tobón C, Köhler L, Schmid S, Burkard R, Frumau KFA, Bruijnzeel LA (2010a) Water dynamics of epiphytic vegetation in a lower montane cloud forest: interception, storage and evaporation of horizontal precipitation. In: Bruijnzeel LA, Scatena FN, Hamilton LS (eds) Tropical montane cloud forests: science for conservation and management. Cambridge University Press, Cambridge, pp 502–515

Tobón C, Bruijnzeel LA, Frumau KFA, Calvo JC (2010b) Changes in soil hydraulic properties and soil water status after conversion of tropical montane cloud forest to pasture in northern Costa Rica. In: Bruijnzeel LA, Scatena FN, Hamilton LS (eds) Tropical montane cloud forests: science for conservation and management. Cambridge University Press, Cambridge, pp 765–778

USDA - Soil Survey Staff (2014) Keys to soil taxonomy, Twelfth Edition, 2014. United States Department of Agriculture Natural Resources Conservation Service. USDA-Natural Resources Conservation Service, Washington, DC. 372 p.

van Dijk IJM, Peña-Arancibia JL, Bruijnzeel LA (2012) Land cover and water yield: inference problems when comparing catchments with mixed land cover. Hydrol Earth Syst Sci 16:461–3473

Vance ED, Nadkarni NM (1990) Microbial biomass and activity in canopy organic matter and the forest floor of a tropical cloud forest. Soil Biol Biochem 22:677–684

Vásquez VG (2016) Influencia del uso de la tierra en la respuesta hidrológica de cuencas de cabecera en los Andes Centrales de Colombia. Disertación doctoral. Tesis, Universidad Nacional de Colombia, Medellín. 146 pp

Veneklaas EJ, Van Ek R (1990) Rainfall interception in two tropical montane rain forests, Colombia. Hydrol Process 4:311–326

Veneklaas EJ, Zagt RJ, van Leerdam A, van Ek R, Broekhoven AJ, van Genderen M (1990) Hydrological properties of the epiphyte mass of a montane tropical rain forest, Colombia. Vegetatio 89:183–192

Vis M (1986) Interception, drop size distributions and rainfall kinetic energy in four Colombian forest ecosystems. Earth Surf Process Landf 11:591–570

Vuille M, Bradley RS, Keimig F (2000) Interannual climate variability in the Central Andes and its relation to tropical Pacific and Atlantic forcing. J Geophys Res 105:12447–12460

Villegas JC, Tobón C, Breshears DD (2008) Fog Interception by non-vascular epiphytes in the tropical cloud forests: dependencies on gauge type and meteorological conditions. Hydrological Processes 22:2484–2492

Walker R, Ataroff M (2005) Intercepción y drenaje en las epífitas de dosel de una selva Nublada andina venezolana. In: Ataroff M, Silva JF (eds) Dinámica Hídrica en Sistemas Neotropicales. ICAE, Univ. Los Andes, Mérida

Walther GR, Post E, Convey P, Menzel A, Parmesan C, Beebee TJC, Fromentin J, Hoegh-Guldberg O, Bairlein F (2002) Ecological responses to recent climate change. Nature 416:389–339

Wassenaar T, Gerber P, Verburg PH, Rosales M, Ibrahim M, Steinfeld H (2007) Projecting land use changes in the Neotropics: the geography of pasture expansion into forest. Glob Environ Chang 17:86–104

Wolf JHD (1993) Diversity patterns and biomass of epiphytic bryophytes and lichens along an altitudinal gradient in the northern Andes. Ann Mo Bot Gard 80:928–960

Wright C, Kagawa-Viviani A, Gerlein-Safdi C, Mosquera G, Poca M, Tseng H, Chun KP (2017) Advancing ecohydrology in the changing tropics: perspectives from early career scientists. Ecohydrology 106(17):e1918. https://doi.org/10.1002/eco.1918

Young KR (2006) Bosques húmedos. In: Moraes R et al (eds) Botánica Económica de los Andes Centrales. Universidad Mayor de San Andrés, La Paz

Chapter 5
Litterfall in Andean Forests: Quantity, Composition, and Environmental Drivers

Wolfgang Wilcke

5.1 Introduction

The carbon storage of an ecosystem is driven by the balance between the production of organic matter and its decomposition. The currently on-going environmental changes in the Andes including increasing temperatures and locally variable positive and negative changes of rainfall (Vuille et al. 2003; Urrutia and Vuille 2009; Peters et al. 2013) and increasing N deposition (Galloway et al. 2004; Wilcke et al. 2013) will likely affect both productivity and decomposition. This together with the increasing atmospheric CO_2 concentrations will possibly change the equilibrium between organic matter production and decomposition. To assess the direction of these changes requires the knowledge of the current productivity and decomposition and their drivers. Net primary production can be estimated with the help of the comparatively easily measured fine litterfall because it was reported that fine litterfall accounted for ca. one-third of total net primary production (Bray and Gorham 1964; Chave et al. 2010). Moreover, litterfall carries nutrients from the forest canopy to the soil and is thus also an important part of the nutrient cycling (Vogt et al. 1986; Veneklaas 1991; Wilcke et al. 2002).

Fine litterfall consists of leaves/needles, twigs, reproductive materials (flowers, fruits, seeds), and other small dead organic material dropping from the forest canopies to the soil and is usually collected with several at least 0.25 m² large traps. To estimate decomposition, e.g., in situ litter bag experiments or the ratio of annual fine litterfall to organic matter storage in the soil organic layer can be used (Vogt et al. 1986).

The comprehensive early review of Vogt et al. (1986) has shown that worldwide litterfall correlates negatively with latitude because it is driven by light availability.

W. Wilcke (✉)
Institute of Geography and Geoecology (IfGG), Karlsruhe Institute of Technology (KIT), Karlsruhe, Germany
e-mail: wolfgang.wilcke@kit.edu

© Springer Nature Switzerland AG 2021
R. W. Myster (ed.), *The Andean Cloud Forest*,
https://doi.org/10.1007/978-3-030-57344-7_5

The correlation is closer for broad-leaved than for needle-leaved forests. The N flux with litterfall in broad-leaved forests (but not in needle-leaved forests) is also negatively correlated with latitude. The N and P fluxes with litterfall are positively correlated in broad-leaved but not in needle-leaved forests. The N concentrations in litterfall decrease from the equator to the pole. The N concentration in litterfall is positively correlated with the mass of litterfall, while there is no consistent relationship between P concentration in litterfall and mass of litterfall. This suggests a higher importance of N than of P availability for litterfall (Vogt et al. 1986).

Several local studies in the humid forests on the Eastern Cordillera in Ecuador and Peru, which forms the rim of the Amazon basin, have found that litterfall changes with topographical elevation. Girardin et al. (2010) observed along an elevational transect from 194 to 3025 m a.s.l. in Peru decreasing litterfall in the order lowland > premontane > upper montane > lower montane forests, but the difference between upper and lower montane forests was small (on average 165 vs. 144 g m^{-2} year^{-1} C, while the lowland forest produced 510 g m^{-2} year^{-1} C). Pinos et al. (2017) determined a surprisingly high leaf litterfall in an upper montane *Polylepis reticulata* Hieron. forest near the treeline at 3735–3930 m a.s.l. in the Cajas National Park in Ecuador of 279 to 465 g m^{-2} year^{-1}. Röderstein et al. (2005) reported that leaf litter mass decreased to less than a third (from 862 to 263 g m^{-2} year^{-1}) with increasing elevation from 1890 m to 3060 m a.s.l. in south Ecuador. Similarly, Wilcke et al. (2008) observed that forest productivity reflected by forest stature, tree basal area, and tree growth decreased from 1960 to 2450 m a.s.l. Wolf et al. (2011) found a maximum of the litterfall at ca. 2000 m a.s.l. along an elevational gradient from 990 to 3000 m a.s.l. again in south Ecuador. In their study, the forests at ca. 2000 m a.s.l. showed a high productivity at the upper end of the range of published litterfall in tropical forests. A similar high litterfall of 850–970 g m^{-2} year^{-1} was reported by Wilcke et al. (2002) for five study sites between 1900 and 2010 m a.s.l. Wolf et al. (2011) furthermore observed that litterfall in the North Andes in south Ecuador decreased from lower slope to upper slope positions at the scale of a few 10s of meters.

Compared with tropical lowland forests and old-world temperate forests, the montane forests of the Andes have been up to now much less studied. Of the 81 study sites included in the comprehensive review of the litterfall in tropical South America (including Panama) of Chave et al. (2010) only five were located in the Andes. The knowledge about litterfall in the southern beech forests (*Nothofagus* sp.) of the south Andes seems even more limited, because I only found a single study from the Chilean Andes including four differently managed and unmanaged stands in the international journal literature (Staelens et al. 2011), although there is some more information about the productivity of *Nothofagus* sp. forests from the south Chilean island of Chiloë outside the Andes (Pérez et al. 1998, 2003).

In this review, I focus on fine litterfall, its fractions (leaves/needles, twigs, reproductive materials, and miscellaneous), its most important macronutrient concentrations N, P, and K, and its C:nutrient ratios along with soil and forest stand properties. My objectives are (1) to compile the published data about litterfall and its fractions in the Andean forests accessible via the Web of Science, (2) to analyze the

5 Litterfall in Andean Forests: Quantity, Composition, and Environmental Drivers

relationship of geographic location (latitude, longitude, elevation) and climate (precipitation, temperature) with litterfall quantity and quality, and (3) to investigate the role of soil, forest stand, and chemical litter quality for litterfall mass.

5.2 Study Sites and Methods

Using the Web of Science in September 2019, I found published reports of litterfall with varying additional information on soil and litter properties from 44 forest stands, which I combined with my own group's results from another 12 sites in a lower montane forest in south Ecuador (Table 5.1). The reviewed studies used at least five samplers per studied forest plot of varying size with a minimum surface area of 0.09 m^2. In some work, only leaf litterfall was reported (León et al. 2011; Pinos et al. 2017), which I extrapolated to total fine litterfall assuming that leaf litterfall accounted for 70% of total litterfall.

The study area in south Ecuador, from which I included results of 12 study sites which are unpublished except for the first measurement year of five of these sites (Wilcke et al. 2002), is located on the eastern slope of the "Cordillera Real," the Eastern Andean Cordillera in south Ecuador facing the Amazon basin at 4° 00'S and 79° 05'W. From 1998 to 2012, 12 measurement sites were run in native old-growth tropical montane rain forest between 1900 and 2130 m a.s.l. for varying durations. One measurement site was located at each of the microcatchments (MC) 1 and 3 at 1900 m a.s.l. (years of 1998–2003), three in MC 2 at 1900, 1950, and 2000 m a.s.l. (1998–2007) (Wilcke et al. 2002), three in MC 5 at 2050–2090 m a.s.l. (Plots F, J, and M, 2004–2009, Wilcke et al. 2009), and four in the Nutrient Management Experiment (NUMEX) between 2060 and 2130 m a.s.l. (the unfertilized control plots of NUMEX, 2007–2009, Homeier et al. 2012).

From 1999 to 2010, annual incident precipitation at the Ecuadorian study site ranged 1900–2700 mm. June was the wettest month with approx. 310 mm and November the driest with approx. 130 mm. Mean annual precipitation is nearly constant between 1950 and 2270 m a.s.l. (Bendix et al. 2008). The mean annual temperature at 1950 m a.s.l. between 1999 and 2010 was 14.9 °C. The coldest month was July, with a mean temperature of 13.7 °C, the warmest November with a mean temperature of 15.8 °C. The average gradient of air temperature between 1950 m a.s.l. and 3180 m a.s.l. in the study area is 0.61 °C/100 m^{-1} (Bendix et al. 2008). Thus, the study area covered an altitudinal gradient of 1.4 °C but received similar annual precipitation.

Recent soils have mainly developed from surface sediments on steep slopes (Wilcke et al. 2001, 2003). The underlying bedrock consists of interbedding of paleozoic phyllites, quartzites, and metasandstones but in the study area phyllites dominate. Dominating soils included Humic Eutrudepts, Humic Dystrudepts, Oxyaquic Eutrudepts, and Histic Humaquepts (Soil Survey Staff 2014). All soils are shallow, loamy-skeletal with high mica contents. The organic layer consisted of Oi,

Table 5.1 Overview of the collected literature results about location, rock and forest type, litterfall, and litterfall components ordered from N to S

	Country	Latitude [°]	Longitude [°]	Elevation [m a.s.l.]	Mean annual precipitation [mm]	Mean annual temperature [°C]	Rock type	Forest type[a]	Litterfall [g m^{-2} year^{-1}]	Leaves (L) [%]	Reproductive materials (R) [%]	R/L	Twigs [%]	Rest [%]	Source
1	Colombia	10.739	−72.833	267	777	27.3		DRY	208	66.6	15.9	0.24	11.1	6.4	Fuentes-Molina et al. (2018)
2	Venezuela	8.617	−71.350	2350	1500	12.6	Schist/ sandstone	UMF	697	48.5	15.6	0.32	32.6	3.3	Steinhardt (1979)
3	Venezuela	8.000	−72.000	2550	2500	13	Gneiss	UMF	430						Tanner et al. (1992)
4	Colombia	6.760	−75.108	1200	2078	23	Granite	PRE	905						Sierra et al. (2007)
5	Colombia	6.541	−75.802	542	1594	26.6	Amphibolite	PLA	56	65.0	24.0	0.37	9.0	2.0	Flórez-Flórez et al. (2013)
6	Colombia	6.300	−75.500	2490	1948	14.9	Amphibolite	UMF	748[b]	71.0			14.3	14.7	León et al. (2011) and Ramírez et al. (2014)
7	Colombia	6.300	−75.500	2490	1948	14.9	Amphibolite	PLA	777[b]	63.0			12.5	24.5	León et al. (2011) and Ramírez et al. (2014)
8	Colombia	6.300	−75.500	2490	1948	14.9	Amphibolite	PLA	349[b]	71.0			20.6	8.4	León et al. (2011) and Ramírez et al. (2014)
9	Colombia	5.000	−75.000	2550	2115	12.2	Volcanic ash	UMF	703	65.6	9.4	0.14	15.1	9.9	Veneklaas (1991)
10	Colombia	5.000	−75.000	3370	1453	7.7	Volcanic ash	UMF	431	65.4	6.3	0.10	17.6	10.7	Veneklaas (1991)

	Country	Latitude [°]	Longitude	Elevation [m a.s.l.]	Mean annual precipitation [mm]	Mean annual temperature [°C]	Rock type	Forest type[a]	Litterfall [g m⁻² year⁻¹]	Leaves (L) [%]	Reproductive materials (R)	R/L	Twigs [%]	Rest	Source
11	Ecuador	−2.773	−79.220	3833	876	5.44	Volcanic ash	UMF	596[b]						Pinos et al. (2017)
12	Ecuador	−2.774	−79.220	3735	876	5.44	Volcanic ash	UMF	397[b]						Pinos et al. (2017)
13	Ecuador	−2.778	−79.226	3890	876	5.44	Volcanic ash	UMF	493[b]						Pinos et al. (2017)
14	Ecuador	−2.780	−79.207	3811	876	5.44	Volcanic ash	UMF	594[b]						Pinos et al. (2017)
15	Ecuador	−2.783	−79.210	3841	876	5.44	Volcanic ash	UMF	484[b]						Pinos et al. (2017)
16	Ecuador	−2.816	−79.223	3930	876	5.44	Volcanic ash	UMF	664[b]						Pinos et al. (2017)
17	Ecuador	−3.967	−79.067	1950	1950	15.7	Schist/ sandstone	LMF	950						Wolf et al. (2011)
18	Ecuador	−3.967	−79.067	1950	1950	15.7	Schist/ sandstone	LMF	980						Wolf et al. (2011)
19	Ecuador	−3.967	−79.067	1950	1950	15.7	Schist/ sandstone	LMF	710						Wolf et al. (2011)
20	Ecuador	−3.975	−79.074	1900	2500	15	Schist/ sandstone	LMF	1084						This study
21	Ecuador	−3.975	−79.074	1900	2500	15	Schist/ sandstone	LMF	1053						This study
22	Ecuador	−3.975	−79.074	1900	2500	15	Schist/ sandstone	LMF	1176						This study
23	Ecuador	−3.975	−79.074	1950	2500	15	Schist/ sandstone	LMF	980						This study

(continued)

Table 5.1 (continued)

	Country	Latitude [°]	Longitude	Elevation [m a.s.l.]	Mean annual precipitation [mm]	Mean annual temperature [°C]	Rock type	Forest type[a]	Litterfall [g m^{-2} year^{-1}]	Leaves (L) [%]	Reproductive materials (R)	R/L	Twigs [%]	Rest	Source
24	Ecuador	−3.975	−79.074	2000	2500	15	Schist/ sandstone	LMF	991						This study
25	Ecuador	−3.975	−79.074	2050	2500	15	Schist/ sandstone	LMF	1285						This study
26	Ecuador	−3.975	−79.074	2050	2500	15	Schist/ sandstone	LMF	1086						This study
27	Ecuador	−3.975	−79.074	2060	2500	15	Schist/ sandstone	LMF	433						This study
28	Ecuador	−3.975	−79.074	2075	2500	15	Schist/ sandstone	LMF	352						This study
29	Ecuador	−3.975	−79.074	2090	2500	15	Schist/ sandstone	LMF	1127						This study
30	Ecuador	−3.975	−79.074	2100	2500	15	Schist/ sandstone	LMF	570						This study
31	Ecuador	−3.975	−79.074	2130	2500	15	Schist/ sandstone	LMF	342						This study
32	Ecuador	−3.976	−79.085	1890	1950	15.7	Schist/ sandstone	LMF	834	59.5	4.3	0.07	10.6	25.7	Moser et al. (2011) and Leuschner et al. (2013)
33	Ecuador	−3.989	−79.082	2380	5000	13.2	Schist/ sandstone	UMF	365	72.3	3.0	0.04	10.7	14.0	Moser et al. (2011) and Leuschner et al. (2013)
34	Ecuador	−4.100	−78.967	1095	2230	19.4	Granodiorite	PRE	700						Wolf et al. (2011)

	Country	Latitude [°]	Longitude	Elevation [m a.s.l.]	Mean annual precipitation [mm]	Mean annual temperature [°C]	Rock type	Forest type[a]	Litterfall [g m^{-2} year^{-1}]	Leaves (L) [%]	Reproductive materials (R)	R/L	Twigs [%]	Rest [%]	Source
35	Ecuador	−4.100	−78.967	1095	2230	19.4	Granodiorite	PRE	860						Wolf et al. (2011)
36	Ecuador	−4.100	−78.967	1095	2230	19.4	Granodiorite	PRE	640						Wolf et al. (2011)
37	Ecuador	−4.100	−79.183	2900	4500	9.4	Schist/sandstone	UMF	470						Wolf et al. 2011
38	Ecuador	−4.100	−79.183	2900	4500	9.4	Schist/sandstone	UMF	480						Wolf et al. (2011)
39	Ecuador	−4.100	−79.183	2900	4500	9.4	Schist/sandstone	UMF	450						Wolf et al. (2011)
40	Ecuador	−4.112	−78.972	1540	2300	17.5	Granodiorite	PRE	734	68.9	5.9	0.08	12.7	12.5	Moser et al. (2011) and Leuschner et al. (2013)
41	Ecuador	−4.115	−78.967	1050	2230	19.4	Granodiorite	PRE	778	64.9	11.7	0.18	14.5	8.9	Moser et al. (2011); Leuschner et al. (2013)
42	Ecuador	−4.120	−79.183	3060	4500	9.4	Schist/sandstone	UMF	270	66.3	2.6	0.04	14.4	16.7	Moser et al. (2011); Leuschner et al. (2013)
43	Peru	−12.959	−71.566	1000	3087	20.7	Alluvial sediments	PRE	476[a]						Girardin et al. (2010) and Zimmermann et al. (2009)

(continued)

Table 5.1 (continued)

	Country	Latitude [°]	Longitude	Elevation [m a.s.l.]	Mean annual precipitation [mm]	Mean annual temperature [°C]	Rock type	Forest type[a]	Litterfall [g m^{-2} year^{-1}]	Leaves (L) [%]	Reproductive materials (R)	R/L	Twigs [%]	Rest	Source
44	Peru	−13.049	−71.537	1500	2631	18.8	Granite	PRE	528[a]						Girardin et al. (2010) and Zimmermann et al. (2009)
45	Peru	−13.071	−71.558	1855	2472	18	Granite	LMF	275[a]						Girardin et al. (2010)
46	Peru	−13.073	−71.559	2020	1827	17.4	Schist	LMF	275[a]						Girardin et al. (2010)
47	Peru	−13.075	−71.589	2720	2318	13.5	Schist	UMF	350[a]						Girardin et al. (2010)
48	Peru	−13.109	−71.600	3020	1776	11.8	Schist	UMF	227[a]						Girardin et al. (2010)
49	Peru	−13.190	−71.587	3025	1706	12.5	Schist	UMF	367	69.0	9.0	0.13	18.0	3.9	Girardin et al. (2010, 2014) and Zimmermann et al. (2009)
50	Peru	−13.190	−71.597	2825	1560	13.1	Schist	UMF	269	73.0	10.0	0.14	16.0	0.9	Girardin et al. (2010)
51	Chile	−39.583	−73.117	805	4500	9	Volcanic ash	SBE	476	66.4	11.1	0.17	21.5	1.0	Staelens et al. (2011)
52	Chile	−39.583	−73.117	734	4500	10	Volcanic ash	SBE	457	81.1	5.6	0.07	11.7	1.6	Staelens et al. (2011)
53	Chile	−39.583	−73.117	630	4500	11	Volcanic ash	SBE	506	76.5	6.6	0.09	16.5	0.3	Staelens et al. (2011)
54	Chile	−39.583	−73.117	637	4500	12	Volcanic ash	SBE	427	69.8	5.6	0.08	24.1	0.5	Staelens et al. (2011)

[a] Litterfall estimated from reported C fluxes with litterfall by assuming a factor of 1.92 taken from the comparison of the report of the C flux with litterfall in Girardin et al. (2010) and the litterfall in Girardin et al. (2014) of one identical site

[b] DRY is dry forest, PLA is plantation, PRE is tropical premontane rain forest, LMF is tropical lower montane forest, UMF is tropical upper montane forest, SBE is southern beech (*Nothofagus* sp.) forest

[c] Litterfall estimated from reported leaf litterfall by assuming a leaf litter contribution of 70%

Oe, and frequently also Oa horizons. The thickness increased with increasing elevation giving Histosols (mainly Terric Haplosaprists) above ca. 2100 m.

The study forest is a Lower Montane Rain Forest (Bruijnzeel and Hamilton 2000). All measurement sites were below old-growth forest. The forest was nearly undisturbed except for MC 5 where in 2004 a slight forest intervention removing 10.2% of the initial basal area was conducted (Günter et al. 2008; Wilcke et al. 2009). More information on the composition of the forest can be found in the work of Homeier (2004).

At each measurement station, litterfall was collected with three to five 0.09 (April 1998–October 2005) or 0.25 m^2 large traps (after October 2005). I analyzed all available monthly litterfall samples from MC1, MC3, and NUMEX, while the data set for MC2 covered April 1998–March 2005 and September 2009–August 2010 and that for MC5 June 2008–May 2009. The thickness of the organic layer was measured on the wall of a soil pit. The lower end of the organic layer was located at the position in the soil profile where the bulk density increased abruptly from >0.3 g cm^{-3} to >1 g cm^{-3}.

Mineral topsoil pH was measured in water (soil:solution ratio 1:2.5 v/v for the mineral soil and 1:10 for the organic horizons) with a glass electrode (Orion U402-S7, Thermo Fisher Scientific, Waltham, MA, USA), total C and N concentrations in soil and litterfall with an Elemental Analyzer (vario EL, Elementar Analysensysteme, Hanau, Germany) on finely ground samples, and effective cation-exchange capacity (ECEC) by extraction with 1 M NH_4NO_3. To determine P, K, Ca, and Mg concentrations in litterfall, finely ground samples were digested with concentrated HNO_3 under pressure (Heinrichs et al. 1986). The concentrations of Al, K, Ca, Mg, and Na were determined with flame atomic absorption spectrometry (AAS; Varian SpectrAA 400) and those of P with inductively-coupled plasma atomic-emission spectrometry (ICP-AES; GBC Integra XMP). Base saturation (BS) was calculated by summing up the proportion of charge equivalents of extractable Ca, K, Mg, and Na and dividing the sum by the ECEC.

5.3 Results

While there is already an ample literature about the litterfall in the tropical forests of the North Andes, the southern part of the Andes (Bolivia, Chile, Argentina) is hardly covered (Table 5.1). I only found reports from four Andean sites under southern beech (*Nothofagus* sp.). The most intensively studied country is Ecuador with 32 sites. Moreover, I only found a single report about litterfall in the dry forests which dominate the inner Andean depressions and the tropical part of the Western Cordillera (Table 5.1; Fuentes-Molina et al. 2018). Most of the available publications referred to old-growth native rain and cloud forests except three which included tree plantations in the rain and cloud forest area (all monocultures; León et al. 2011; Flórez-Flórez et al. 2013).

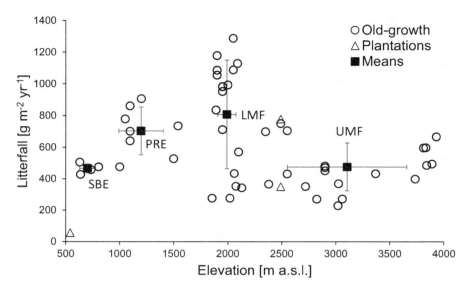

Fig. 5.1 Relationship between elevation and litterfall in 50 old-growth native rain forest stands (open circles) and three 5–43-year-old tree plantations (open triangles); a 5-year-old *Azadirachta indica* A.Juss., a 43-year-old *Pinus patula* Schltdl. & Cham. and a 43-year-old *Cupressus lusitanica* Mill. plantation in the Andes (for the data sources see Table 1). Black rectangles illustrate mean litterfall of southern beech (SBE), premontane (PRE), lower montane (LMF), and upper montane forests (UMF)

Litterfall of the old-growth native rain forests increased from higher and lower latitudes to the equator and was independent from longitudinal location (Table 5.1). In these old-growth native rain and cloud forests, litterfall showed a hump-shaped elevational distribution with maximal productivity around 2000 m a.s.l. (Fig. 5.1). The litterfall of the two 43-year-old tree plantations (León et al. 2011) fell well within the range of litterfall of the old-growth forests, while the 5-year-old tree plantation (Flórez-Flórez et al. 2013) showed the lowest litterfall of the whole data set (triangles in Fig. 5.1). When grouped into forest types, mean litterfall increased from premontane (1000–1540 m a.s.l., $n = 8$) to lower montane forests (including lower montane rain and cloud forests; Bruijnzeel and Hamilton 2000; 1855–2130 m a.s.l., $n = 18$) and decreased to upper montane forests (including upper montane cloud and stunted subalpine forests/elfin forests; Bruijnzeel and Hamilton 2000; Bruijnzeel et al. 2011; 2350–3950 a.s.l., $n = 20$). The given elevational range refers to the particular study sites included in this review and should not be considered as general elevation ranges of the considered forest types, particularly because these elevational ranges can locally vary. The temperate southern beech forests (630–805, $n = 4$) showed a similar productivity as the tropical upper montane forests (Fig. 5.1). At the same time the variation in litterfall was highest in the lower montane forest ranging from 270 to 1280 g m^{-2} year^{-1} with a coefficient of variation of 42% as compared to 21%, 32%, and 6% for premontane, upper montane, and southern beech forests, respectively. The four southern beech forests showed the smallest variation in litterfall in spite of the fact that these stands included variable forest properties such as evergreen and deciduous and native, modified, and managed stands (Staelens et al. 2011).

Litterfall consisted of 68 ± s.d. 7.2% leaves, 17 ± 5.6% woody material, 7.6 ± 3.5% reproductive material, and 8.3 ± 7.3% unidentifiable rest ($n = 15$; Table 5.1). The composition of the litterfall did not show a discernible spatial pattern, which was also true for the ratio of reproductive material to leaves in litterfall (0.12 ± 0.07). However, the 5-year-old tree plantation produced with 24% a higher contribution of reproductive materials than the old-growth native rain forests, followed with 16% by the dry forest (Table 5.1). There were no data on the contribution of reproductive materials to litterfall for the other two 43-year-old tree plantations. The mineral topsoils were consistently acidic (Table 5.2). The pH of the mineral topsoil did not change systematically with elevation and was not related with litterfall. The mineral topsoil of the 5-year-old tree plantation was the only one with a slightly alkaline pH of 7.3 (Flórez-Flórez et al. 2013). There was also no relationship between clay concentration, effective cation-exchange capacity, base saturation, and N concentration of the mineral topsoil with elevation or litterfall. Organic layer thickness correlated positively with elevation ($r = 0.57$, $p < 0.001$, $n = 34$) and litterfall correlated negatively with organic layer thickness (Fig. 5.2) in the tropical old-growth native rain and cloud forests. There were no data of organic layer thickness in the dry and southern beech forests and in the plantations. The C/N ratios of the mineral soil correlated negatively with litterfall ($r = 0.70$, $p < 0.001$, $n = 23$).

Mean stem density was highest in the upper montane forest and similar in the three other forest types (Table 5.3). Mean basal area and mean tree height were highest in southern beech forests. In the tropical rain forests, mean basal area was similar and mean tree height decreased from premontane to lower montane forests but did not change with further increasing elevation. Leaf area index (LAI) increased from premontane to lower montane and decreased to upper montane forests in line with litterfall although LAI did not correlate with litterfall, when all 18 individual study sites, for which LAI data were available, were considered. The study of litterfall in the southern beech forests did not report LAI (Staelens et al. 2011). There was also no relationship between the number of stems with a dbh > 10 cm ha^{-1}, basal area, and mean tree height with litterfall.

The N and P concentrations in litterfall tended to be lowest in the upper montane < southern beech < premontane ≈ lower montane forests (Table 5.3). The K concentrations were lowest in the southern beech forests < upper montane < lower montane, while there were no K data for premontane forests. There was no relationship between the N, P, and K concentrations in litterfall of the old-growth rain and cloud forests and elevation. However, C/N ($n = 25$) and C/P ratios decreased marginally significantly ($p < 0.1$) with increasing elevation, while N/P ratios ($p = 0.437$) did not change (Fig. 5.3). There was also no relationship between elevation and N/K and P/K ratios. There were significant positive correlations of N ($r = 0.82$, $p < 0.001$,

Table 5.2 Mean soil properties (standard deviations; number of measurements) and pH ranges (number of measurements) of the four old-growth native rain forest types. For the data sources see Table 5.1

Forest type	Organic layer thickness [mm]	pH in H$_2$O[a]	Sand [%]	Silt	Clay	Cation-exchange capacity [nmol$_c$ g^{-1}]	Base saturation [%]	C	N	P	C/N concentration ratio
								concentration [mg g^{-1}]			
Premontane	81 (76; 7)	3.9–5.1 (7)	49 (19; 5)	41 (17; 5)	19 (16; 5)	No data	43 (35; 3)	180 (100; 2)	10 (4; 2)	0.70 (0.10; 3)	18 (2.9; 5)
Lower montane	18 (110; 18)	3.6–5.3 (13)	31 (6.6; 7)	52 (4.9; 7)	15 (3.4; 7)	66 (42; 9)	31 (36; 12)	29 (19; 12)	2.4 (1.7; 9)	0.62 (0.19; 8)	16 (2.0; 12)
Upper montane	270 (140; 9)	3.3–6.3 (11)	25 (11; 4)	57 (10; 4)	17 (2.9; 4)	148 (−; 1)	22 (−; 1)	84 (55; 3)	4.9 (2.7; 3)	0.22 (0.20; 5)	18 (3.7; 6)
Southern beech	No data	4.7–5.4 (4)	No data	No data	No data	No data	No data	140 (190; 4)	No data	No data	No data

[a]pH measurements in CaCl$_2$ were converted to pH in H$_2$O by multiplying with 1.15 based on a regression of pH(H$_2$O) on pH(CaCl$_2$) ($r = 0.94$, $p = 0.019$, $n = 5$)

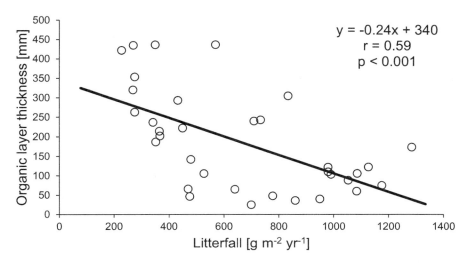

Fig. 5.2 Relationship between litterfall and organic layer thickness in the old-growth native rain and cloud forests ($n = 34$)

$n = 30$), P ($r = 0.65$, $p < 0.001$, $n = 30$), and K concentrations ($r = 0.75$, $p < 0.001$, $n = 20$) in litterfall with annual litterfall mass.

As a consequence of the small variation in nutrient concentrations, the fluxes of N, P, and K with litterfall in the tropical rain forests followed the same elevational distribution as litterfall, i.e., they showed a hump-shaped elevational distribution (Fig. 5.1; Table 5.3). The N and P fluxes with litterfall in the southern beech forests were again similar to those in the upper montane forests, but the K fluxes were considerably lower. The fluxes of N with litterfall correlated closely with those of P and K (Fig. 5.4).

5.4 Discussion

5.4.1 Litterfall and Litterfall Fractions in Andean Forests

Based on the available data, it is hardly possible to draw a consistent and comprehensive picture of litterfall in the Andes, because there are large regions that are almost not covered by data including Bolivia, Chile, and Argentina (Table 5.1). Moreover, I only found one reference reporting litterfall in dry forests (Fuentes-Molina et al. 2018), which is with ca. 200 g m^{-2} year^{-1} lower than in all other forest types. Dry forests cover large parts of the Pacific-exposed slope and the inner Andean depressions in the tropical part of the Andes. The Tumbesian dry forest in Ecuador and Peru as an example, which has been comparatively intensively studied, covers 87,000 km^2 (Espinosa et al. 2011). Therefore, the productivity of the Andean

Table 5.3 Mean forest stand properties, chemical litter quality, and nutrient fluxes (standard deviations; number of measurements) of the four old-growth native rain and cloud forest types. For the data sources see Table 5.1

Forest type	Number of trees >10 cm dbh [ha⁻¹]	Basal area [m² ha⁻¹]	Mean tree height [m]	Leaf area index [m² m⁻²]	C	N	P	K	C/N	C/P	N/P	N	P	K
					Concentration [mg g⁻¹]							Flux [g m⁻² year⁻¹]		
Premontane	1080 (624; 5)	36.7 (14.6; 5)	26.8 (7.14; 2)	5.70 (0.42; 2)	530 (159; 3)	15.9 (0.60; 3)	0.77 (0.06; 3)	No data	33.4 (2.6; 3)	695 (76.2; 3)	39.5 (1.90; 3)	11.7 (1.94; 3)	0.57 (0.12; 3)	No data
Lower montane	1010 (418; 14)	32.7 (9.34; 14)	15.4 (2.99; 10)	6.9 (0.79; 6)	468 (73.8, 15)	15.3 (4.08; 15)	0.85 (0.45; 15)	6.23 (2.63; 12)	33.3 (11.6; 15)	725 (387; 15)	39.8 (13.6; 15)	14.4 (7.08; 15)	0.82 (0.55; 15)	6.11 (3.78; 12)
Upper montane	2730 (2560; 15)	41.0 (16.5; 15)	16.8 (5.78; 8)	3.70 (1.19; 10)	543 (14.9; 3)	9.86 (1.84; 8)	0.50 (0.19; 8)	4.37 (3.04; 4)	59.1 (17.3; 3)	1360 (484; 3)	43.3 (4.73; 3)	5.54 (1.83; 8)	0.28 (0.15; 8)	2.89 (2.27; 4)
Southern beech	1180 (677; 4)	75.7 (38.0; 4)	34.5 (9.53; 4)	No data	448 (18.4;4)	12.2 (0.08; 4)	0.61 (0.07; 4)	2.84 (2.35; 4)	36.6 (1.43; 4)	744 (68.6; 4)	38.7 (3.96; 4)	5.93 (1.02; 4)	0.29 (0.06; 4)	1.45 (1.22; 4)

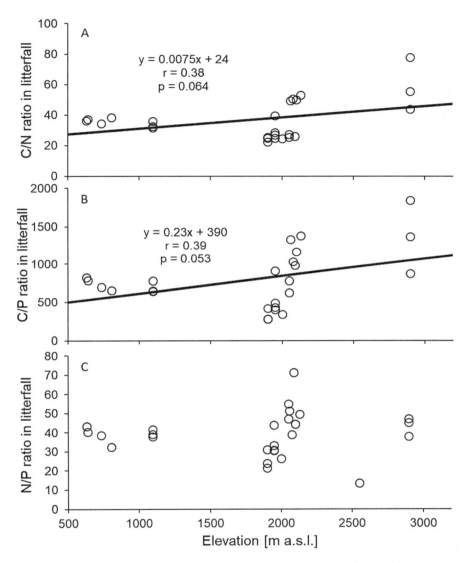

Fig. 5.3 Relationship between elevation and (**a**) C/N, (**b**) C/P, and (**c**) N/P concentration ratios in litterfall of old-growth rain and cloud forests ($n = 25$ for C/N and C/P and $n = 27$ for N/P)

dry forests and their response to changing environmental conditions cannot be neglected when considering the C budget and the mountain risks of the Andes, which depend on the forest cover. Consequently, more work about the productivity of the Andean dry forests is needed.

The three studies on litterfall in plantations (Léon et al. 2011; Flórez-Flórez et al. 2013) suggest that the plantations produce considerably less litterfall in the first few

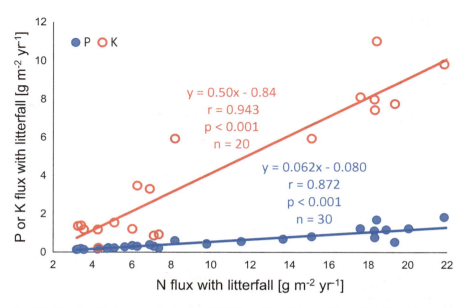

Fig. 5.4 Relationship between the fluxes of N in litterfall and those of P and K in litterfall

years but can reach a similar productivity as the native forests in a few decades. This might indicate that the afforestation of the wide-spread abandoned low-fertility pastures is a promising way to restore the C stocks and budget of the native forest and to reduce mountain risks, including enhanced erosion and less buffered water levels of the headwater streams. However, currently most of the plantations are monocultures of exotic species including those of the genera *Pinus*, *Cupressus*, and *Eucalyptus*. There have only been few attempts to establish plantations of native species, mostly because of limited silvicultural knowledge of these species (Weber et al. 2008). To avoid ecological problems associated with the establishment of monocultures of exotic tree species, more research into the silvicultural management and the economic value chains of native tree species with a potential to be grown in plantations, possibly even in mixed-species stands is needed, if ecosystem C storage and biomass productivity on degraded lands are to be restored in a sustainable way.

The premontane forests had on average a similar litterfall as the Andean premontane forest included in the early evaluation of Vitousek (1984) of 780 g m^{-2} year^{-1} and the range of the lower montane forests was with 275–1285 g m^{-2} year^{-1} slightly wider than spanned by the two Andean lower montane forests in Vitousek (1984) with 780–1160 g m^{-2} year^{-1} (Table 5.1; Fig. 5.1). The Andean forests with the highest productivity, the tropical lower montane forests of the Eastern Cordillera of the North Andes, showed an only slightly lower mean productivity as the old-growth tropical lowland forests in South America (Table 5.1; 861 ± 191 g m^{-2} year^{-1}; Chave et al. 2010). The Andean southern beech forests had a higher litterfall than a broadleaved *Nothofagus*-dominated and a coniferous *Fitzroya*-dominated forest of the

Chilean island of Chiloë with 304 ± 62 and 174 ± 69 g m^{-2} year^{-1} of litterfall, respectively (Pérez et al. 1998, 2003).

The mean contribution of leaf litterfall to total litterfall in the data set evaluated here is close to the mean value of $70.8 \pm 8.5\%$ of the data set from tropical South America of Chave et al. (2010) suggesting that the contribution of leaf litterfall to total litterfall varies little in old-growth forests (Table 5.1). The same seems true for the contribution of reproductive materials to total litterfall for which Chave et al. (2010) reported $8.9 \pm 5.6\%$. Consequently, the mean R/L ratio of 0.118 ± 0.071 in my data set was also similar to that in the forests of tropical South America studied by Chave et al. (2010) with a mean R/L ratio of 0.135 ± 0.119. However, Chave et al. (2010) pointed at the fact that this R/L ratio may be an underestimation, because concerted fruiting during mast years might not be appropriately included in this measure, which is also the case in my data set.

5.4.2   Relationship of Geographical Location and Climate with Litterfall

The litterfall increased towards the equator as already reported by Bray and Gorham (1964) and Vogt et al. (1986) because Jordan (1971) had shown that light availability during the growing season, which increases from the higher latitudes to the equator, was strongly correlated with litterfall (Table 5.1). Vogt et al. (1986) found that particularly broad-leaved forests like most of the here included forests showed a close correlation with latitude, while needle-leaved forests correlated less strongly with latitude. There was no relationship between longitude and litterfall likely owing to the small longitudinal spread of the Andes and possibly also because longitude is confounded with elevation. The hump-shaped elevational distribution of litterfall in the humid tropical part of the Andes with a maximum around 2000 m a.s.l. was also reported by Wolf et al. (2011) from Ecuador (Fig. 5.1). The data of the study of Wolf et al. (2011) is included here (Table 5.1).

The finding that there was no relationship between rainfall and litterfall was also reported in the review of Chave et al. (2010) covering tropical South America, which included five tropical montane forest sites, the data of which are also included here. Although increasing temperature generally stimulates biological activity and thus should increase litterfall as was observed by Vogt et al. (1986) for a global data set, temperature and litterfall were unrelated in the Andes. This finding might be attributable to the fact that in a mountainous area temperature and precipitation interact to influence nutrient supply and water stress in a complex way.

5.4.3 Role of Soil, Forest Stand, and Chemical Litter Quality for Litterfall Mass

The lacking relationship between the pH of the mineral topsoil and litterfall can be attributed to the small variation in pH values, which fell mostly in the aluminum oxide buffer range (Table 5.2), except for the soil under the young plantation. Similarly, Schawe et al. (2007) observed no consistent change in soil pH along an elevational gradient in Peru from 1700 to 3300 m a.s.l. In contrast, Wilcke et al. (2008) reported that pH of A and B horizons decreased with increasing elevation between 1960 and 2450 m a.s.l. in Ecuador. The variation in other soil properties such as texture, cation-exchange capacity, and base saturation, for which again no relationship with litterfall was detected, was similarly small in the included studies.

Organic layer thickness correlated negatively with litterfall although increasing litterfall provides more substrate for the formation of an organic layer on top of the mineral soil. This illustrates that other factors than litterfall are more important for the thickness of the organic layer (Fig. 5.2) including other organic matter sources such as root litter and abiotic conditions (climate, nutrient supply). Vogt et al. (1986) reported in their review that there was no general relationship between litterfall and organic layer thickness and attributed this mainly to the role of root litter. Leuschner et al. (2013) have shown that total belowground biomass increased with increasing elevation under humid tropical forest in Ecuador from 1050 to 3060 m a.s.l., while aboveground biomass decreased. For a partly overlapping transect from 1960 to 2450 m a.s.l. in Ecuador, Wilcke et al. (2008) observed that the nutrient concentrations in O and A horizons decreased with increasing elevation resulting in decreasing nutrient supply which hampers organic matter degradation and results in increasing organic layer thickness.

The finding that none of the forest stand properties listed in Table 5.3 correlated with litterfall is unexpected. In the literature, it was, for instance, shown for several coniferous forest plantations that the tree basal area correlated with litterfall (Bueis et al. 2017; Erkan et al. 2018). Apparently, the static forest stand properties do not sufficiently reflect the dynamics of the litterfall in the structurally and species-rich old-growth Andean forests.

The finding that the C/N and C/P ratios in litterfall increased with increasing elevation (Fig. 5.3; Table 5.3) confirms an increasing nutrient-use efficiency already reported in Vitousek (1984). However, the fact that no ratio of the nutrient elements (i.e., N/P, N/K, and P/K) showed a systematic change with elevation suggests that there is no general change in the kind of nutrient limitation along the considered elevational range, although it must be borne in mind that all three major nutrients can be retranslocated to variable degrees during leaf senescence (Marschner 2012). In the literature, contrasting results with respect to the relationship between elevation and nutrient limitation have been reported. The classical view based on the work of Vitousek (1984) is that lowland forests are mainly P and montane forests mainly N-limited. Fisher et al. (2013), however, reported an N + P limitation of lowland forests in Peru, while the Andean forests were more N-limited and particularly so with increasing elevation. Similarly, Tanner et al. (1992) found a positive response of litterfall to N addition in a Venezuelan tropical montane forest confirming the classical view. In contrast, Homeier et al. (2012, 2013) reported a significant increase in litterfall after low-level addition of N + P at 2000 m a.s.l. in an Ecuadorian lower montane rain forest after only one year and also found the strongest response of litterfall to fertilization with N + P at 3000 m.s.l., while litterfall in the premontane forest at 1000 m a.s.l. hardly responded to any nutrient addition. Thus, the kind of nutrient limitation seems to vary among different locations and as a consequence no general kind of nutrient limitation appears in the overall data set. The lack of a general kind of nutrient limitation is also reflected by the fact that the fluxes of all considered nutrients correlated closely with each other illustrating that the concentrations of all nutrients in litterfall responded in a similar way (Fig. 5.4). However, similar to the findings of Vogt et al. (1986) for N and P, the correlations between the concentrations of P and K in litterfall and litterfall mass were slightly less close than that of N in litterfall with litterfall mass, supporting a generally more important role of N for litterfall than of P and K.

5.5 Conclusions

The currently available litterfall data from Andean forests has major gaps. Particularly, the south Andes including Bolivia, Chile, and Argentina are hardly covered. Moreover, there is almost no litterfall data from the dry forest, which covers large parts of the inner Andes and there is little data about the litterfall in the far-spread tree plantations of mostly exotic *Pinus*, *Cupressus*, and *Eucalyptus* species. The composition of litterfall showed little variation with ca. 70% leaf litterfall and ca. 10% flower and fruit litterfall.

Litterfall increased towards the equator and showed a hump-shaped elevational distribution. The most productive forests of the Andes are the lower montane forests in the tropical north part. The temperate forest in the south Andes has a similar productivity as the tropical upper montane forests. Litterfall and organic layer thickness are negatively related illustrating different controls of litterfall and organic layer thickness.

Forests became increasingly nutrient-efficient with increasing elevation but there was no general change in nutrient relationships suggesting that the kind of nutrient limitation varies among different locations. Nevertheless, concentrations of N in litterfall showed a closer correlation with litterfall mass than those of P and K confirming a generally slightly more important role of N for the productivity of Andean forests than of other nutrients.

The observed elevational influence of litterfall in the humid tropical Andes suggests that the forest productivity will likely respond to climate change driving the vegetation belts to higher elevation with an unknown overall effect on C sequestration by net primary production of these forests.

108 W. Wilcke

Acknowledgments I thank E. Beck, K. Müller-Hohenstein, M. Richter, and W. Zech for co-initiating the long-term study; C. Valarezo and the National University of Loja for their long-term cooperation, J. Boy, K. Fleischbein, R. Goller, M. Meyer-Grünefeldt, M. Sequeira, A. Velescu, H. Wullaert, S. Yasin, and numerous Ecuadorian and German graduate and undergraduate students for data acquisition during parts of the observation period; the Ecuadorian Environmental Ministry for the research permits; Naturaleza y Cultura Internacional (NCI) in Loja for providing the study area and the research station; and the Deutsche Forschungsgemeinschaft (DFG) for funding (FOR 402 and 816).

References

Bray JR, Gorham E (1964) Litter production in forests of the world, Adv Ecol Res 2:101–157

Bendix J, Rollenbeck R, Richter M, Fabian P, Emck P (2008) Chapter 8: Climate. In: Beck E, Bendix J, Kottke I, Makeschin F, Mosandl R (eds) Gradients in a Tropical Mountain Ecosystem of Ecuador. Ecol Stud 198, pp 63–73, Springer-Verlag, Heidelberg

Bruijnzeel LA, Hamilton LS (2000) Decision time for cloud forests. In: IHP Humid Tropics Programme Series 13, IHP-UNESCO, Paris

Bruijnzeel LA, Mulligan M, Scatena FM (2011) Hydrometeorology of tropical montane cloud forests: emerging patterns. Hydrol Process 25:465–498

Bueis T, Bravo F, Pando V, Belén-Turrión M (2017) Influencia de la densidad del arbolado sobre el desfronde y su reciclado en pinares de repoblación del norte de España. Bosque 38:401–407

Chave J, Navarrete D, Almeida S, Álvarez E, Aragão LEOC, Bonal D, Châtelet P, Silva -Espejo JE, Goret J-Y, von Hildebrand P, Jiménez E, Patiño S, Peñuela MC, Pillips OL, Stevenson P, Malhi Y (2010) Regional and seasonal patterns of litterfall in tropical South America. Biogeosciences 7:43–55

Erkan N, Comez A, Aydin AC, Denli O, Erkan S (2018) Litterfall in relation to stand parameters and climatic factors in *Pinus brutia* forests in Turkey. Scand J For Res 33:338–346

Espinosa CI, Cabrera O, Luzuriaga AL, Escudero A (2011) What factors affect diversity and species composition of endangered Tumbesian dry forests in southern Ecuador? Biotropica 43:15–22

Fisher JB, Malhi Y, Cuba Torres I, Metcalfe DB, van de Weg MJ, Meir P, Silva-Espejo JE, Huaraca Huasco W (2013) Nutrient limitation in rainforests and cloud forests along a 3,000-m elevation gradient in the Peruvian Andes. Oecologia 172:889–902

Flórez-Flórez CP, León-Peláez JD, Osorio-Vega NW, Restrepo-Llano MF (2013) Dinámica de nutrientes en plantaciones forestales de *Azadirachta indica* (Meliaceae) establecidas para restauración de tierras degradadas en Colombia. Rev Biol Trop 61:515–529

Fuentes-Molina N, Rodriguez-Barrios J, Isenia-Leon S (2018) Caída y descomposición de hojarasca en los bosques ribereños del Manantial de Cañaverales, Guajira, Colombia. Acta Biol Colomb 23:115–123

Galloway JN, Dentener FJ, Capone DG, Boyer EW, Howarth RW, Seitzinger SP, Asner GP, Cleveland CC, Green PA, Holland EA, Karl DM, Michaels AF, Porter JH, Townsend AR, Vörösmarty CJ (2004) Nitrogen cycles: past, present, and future. Biogeochemistry 70:153–226

Girardin CAJ, Malhi Y, Aragão LEOC, Mamani-Solórzano M, Huaraca-Huasco W, Durand L, Feeley KJ, Rapp J, Silva-Espejo JE, Silman MR, Salinas N, Whittaker RJ (2010) Net primary productivity allocation and cycling of carbon along a tropical forest elevational transect in the Peruvian Andes. Glob Change Biol 16:3176–3192

Girardin CAJ, Silva-Espejo JE, Doughty CE, Huaraca-Huasco W, Metcalfe DB, Durand-Baca L, Marthews TR, Aragão LEOC, Farfán-Rios W, García-Cabrera K, Halladay K, Fisher JB, Galiano-Cabrera DF, Huaraca-Quispe LP, Alzamora-Taype I, Eguiluz-Mora L, Salinas -Revilla

N, Silman MR, Meir P, Malhi Y (2014) Productivity and carbon allocation in a tropical montane cloud forest in the Peruvian Andes. Plant Ecol Div 7:107–123

Günter S, Cabrera O, Weber M, Stimm B, Zimmermann M, Fiedler K, Knuth J, Boy J, Wilcke W, Iost S, Makeschin F, Werner F, Gradstein R, Mosandl R (2008) Chapter 26: Natural forest management in neotropical mountain rain forests – an ecological experiment. In: Beck E, Bendix J, Kottke I, Makeschin F, Mosandl R (eds) Gradients in a Tropical Mountain Ecosystem of Ecuador. Ecol Stud 198, pp 347–359, Springer-Verlag, Heidelberg

Heinrichs H, Brumsack H-J, Loftfield N, König N (1986) Verbessertes Druckaufschlußsystem für biologische und anorganische Materialien. Z Pflanzenernaehr Bodenkd 149:350–353

Homeier J (2004) Baumdiversität, Waldstruktur und Wachstumsdynamik zweier tropischer Bergregenwälder in Ecuador und Costa Rica. Dissertationes Botanicae 391. J. Cramer, Berlin

Homeier J, Hertel D, Camenzind T, Cumbicus NL, Maraun M, Martinson GO, Poma LN, Rillig MC, Sandmann D, Scheu S, Veldkamp E, Wilcke W, Wullaert H, Leuschner C (2012) Tropical Andean forests are highly susceptible to nutrient inputs – rapid effects of experimental N and P addition to an Ecuadorian montane forest. PLoS One 7:e47128

Homeier J, Leuschner C, Bräuning A, Cumbicus NL, Hertel D, Martinson GO, Spannl S, Veldkamp E (2013) Chapter 23: Effects of nutrient addition on the productivity of montane forests and implications for the carbon cycle. In: Bendix J, Beck E, Bräuning A, Makeschin F, Mosandl R, Scheu S, Wilcke W (eds) Ecosystem Services, Biodiversity and Environmental Change in a Tropical Mountain Ecosystem of South Ecuador. Ecol Stud 221, pp 315–329, Springer-Verlag, Berlin

Jordan CF (1971) A world pattern in plant energetics. Am Sci 59:426–433

León JD, González MI, Gallardo JF (2011) Ciclos biogeoquímicos en bosques naturales y plantaciones de coníferas en ecosistemas de alta montaña de Colombia. Rev Biol Trop 59:1883–1894

Leuschner C, Zach A, Moser G, Homeier J, Graefe S, Hertel D, Wittich B, Soethe N, Iost S, Röderstein M, Horna V, Wolf K (2013) Chapter 10: The carbon balance of tropical mountain forests along an altitudinal transect. In: Bendix J, Beck E, Bräuning A, Makeschin F, Mosandl R, Scheu S, Wilcke W (eds) Ecosystem Services, Biodiversity and Environmental Change in a Tropical Mountain Ecosystem of South Ecuador. Ecol Stud 221, pp 117–139, Springer-Verlag, Berlin

Marschner P (ed) (2012) Marschner's Mineral Nutrition of Higher Plants, 3rd edn. Academic Press/Elsevier, Amsterdam

Moser G, Leuschner C, Hertel D, Graefe S, Soethe N, Iost S (2011) Elevation effects on the carbon budget of tropical mountain forests (S Ecuador): the role of the belowground compartment. Glob Change Biol 17:2211–2226

Pérez CA, Hedin LO, Armesto JJ (1998) Nitrogen mineralization in two unpolluted old-growth forests of contrasting biodiversity and dynamics. Ecosystems 1:361–373

Pérez CA, Armesto JJ, Torrealba C, Carmona MR (2003) Litterfall dynamics and nitrogen use efficiency in two evergreen temperate rainforests of southern Chile. Austral Ecol 28:591–600

Peters T, Drobnik T, Meyer H, Rankl M, Richter M, Rollenbeck R, Thies B, Bendix J (2013) Chapter 2: Environmental changes affecting the Andes of Ecuador. In: Bendix J, Beck E, Bräuning A, Makeschin F, Mosandl R, Scheu S, Wilcke W (eds) Ecosystem Services, Biodiversity and Environmental Change in a Tropical Mountain Ecosystem of South Ecuador. Ecol Stud 221, pp 19-29, Springer-Verlag, Berlin

Pinos J, Studholme A, Carabajo A, Gracia C (2017) Leaf litterfall and decomposition of *Polylepis reticulata* in the treeline of the Ecuadorian Andes. Mount Res Develop 37:87–96

Ramírez JA, León-Peláez JD, Craven D, Herrera DA, Zapata CM, González-Hernández MI, Gallardo-Lancho J, Osorio W (2014) Effects on nutrient cycling of conifer restoration in a degraded tropical montane forest. Plant Soil 378:215–226

Röderstein M, Hertel D, Leuschner C (2005) Above- and belowground litter production in three topical montane forests in southern Ecuador. J Trop Ecol 21:483–492

Schawe M, Glatzel S, Gerold G (2007) Soil development along an altitudinal transect in a Bolivian tropical montane rainforest: Podzolization vs. hydromorphy. Catena 69:83–90

Sierra CA, Harmon ME, Moreno FH, Orrego SA, Del Valle JI (2007) Spatial and temporal variability of net ecosystem production in a tropical forest: testing the hypothesis of a significant carbon sink. Glob Change Biol 13:838–853

Soil Survey Staff (2014): Keys to Soil Taxonomy, 12th ed. United States Department of Agriculture, Natural Resources Conservation Service, Washington DC, USA

Staelens J, Ameloot N, Amonacid L, Padilla E, Boeckx P, Huygens D, Verheyen K, Oyarzún C, Godoy R (2011) Litterfall, litter decomposition and nitrogen mineralization in old-growth evergreen and secondary deciduous *Nothofagus* forests in south-Central Chile. Rev Chil Hist Nat 84:125–141

Steinhardt U (1979) Untersuchungen über den Wasser- und Nährstoffhaushalt eines andinen Wolkenwaldes in Venezuela. In: Göttinger Bodenkundliche Berichte, vol 56. University of Göttingen, Göttingen, pp 1–146

Tanner EVJ, Kapos V, Franco W (1992) Nitrogen and phosphorus fertilization effects on Venezuelan montane forest trunk growth and litterfall. Ecology 73:78–86

Urrutia R, Vuille M (2009) Climate change projections for the tropical Andes using a regional climate model: temperature and precipitation simulations for the end of the 21st century. J Geophys Res 114:D02108

Veneklaas EJ (1991) Litterfall and nutrient fluxes in two montane tropical rain forests, Colombia. J Trop Ecol 7:319–336

Vitousek PM (1984) Litterfall, nutrient cycling, and nutrient limitation in tropical forests. Ecology 65:285–298

Vogt KA, Grier CC, Vogt DJ (1986) Production, turnover, and nutrient dynamics of above- and belowground detritus of world forests. Adv Ecol Res 15:303–377

Vuille M, Bradley RS, Werner M, Keimig F (2003) 20th century climate change in the tropical Andes: observations and model results. Clim Chang 59:75–99

Weber M, Günter S, Aguirre N, Stimm B, Mosandl R (2008) Chapter 34: Reforestation of abandoned pastures: silvicultural means to accelerate forest recovery and biodiversity. In: Beck E, Bendix J, Kottke I, Makeschin F, Mosandl R (eds) Gradients in a tropical mountain ecosystem of Ecuador. Ecol Stud 198, pp 431–441, Springer-Verlag, Heidelberg

Wilcke W, Yasin S, Valarezo C, Zech W (2001) Change in water quality during the passage through a tropical montane rain forest in Ecuador. Biogeochemistry 55:45–72

Wilcke W, Yasin S, Abramowski U, Valarezo C, Zech W (2002) Nutrient storage and turnover in organic layers under tropical montane rain forest in Ecuador. Eur J Soil Sci 53:15–27

Wilcke W, Valladarez H, Stoyan R, Yasin S, Valarezo C, Zech W (2003) Soil properties on a chronosequence of landslides in montane rain forest, Ecuador. Catena 53:79–95

Wilcke W, Oelmann Y, Schmitt A, Valarezo C, Zech W, Homeier J (2008) Soil properties and tree growth along an altitudinal transect in Ecuadorian tropical montane forest. J Plant Nutr Soil Sci 171:220–230

Wilcke W, Günter S, Alt F, Geißler C, Boy J, Knuth J, Oelmann Y, Weber M, Valarezo C, Mosandl R (2009) Response of water and nutrient fluxes to improvement fellings in a tropical montane forest in Ecuador. For Ecol Manag 257:1292–1304

Wilcke W, Leimer S, Peters T, Emck P, Rollenbeck R, Trachte K, Valarezo C, Bendix J (2013) The nitrogen cycle of tropical montane forest in Ecuador turns inorganic under environmental change. Glob Biogeochem Cycle 27:1194–1120

Wolf K, Veldkamp E, Homeier J, Martinson GO (2011) Nitrogen availability links forest productivity, soil nitrous oxide and nitric oxide fluxes of a tropical montane forest in southern Ecuador. Glob Biogeochem Cycle 25:GB4009

Zimmermann M, Meir P, Bird MI, Malhi Y, Ccahuana AJQ (2009) Climate dependence of heterotrophic soil respiration from a soil-translocation experiment along a 3000 m tropical forest altitudinal gradient. Eur J Soil Sci 60:895–906

Chapter 6
Arbuscular Mycorrhizal Fungi and Ectomycorrhizas in the Andean Cloud Forest of South Ecuador

Ingeborg Haug, Sabrina Setaro, and Juan Pablo Suárez

6.1 Introduction

First recognized by A. B. Frank (1885) in forests around Berlin and later corroborated by thousands of scientific studies, it was shown that most plants depend on fungi for nutrient uptake and—by a symbiotic interaction called mycorrhiza—support these root-inhabiting fungi by delivering carbohydrates and amino acids. While ectomycorrhizas in the boreal forests dominate with the fungal partners Ascomycetes and Basidiomycetes, in tropical forests most trees have arbuscular mycorrhizas with Glomeromycotina. The detection of ectomycorrhizal fungi is standardized, while the description of environmental communities of arbuscular mycorrhizal fungi at the species level remains challenging: no sexual structures of the Glomeromycotina are known, thus the microscopic spores are the only morphological structures to describe species. Isolation of spores—especially from humus soil—is difficult, and reliable determinations can only be done by specialists. Added to this is the problem that not all AM fungi form spores in equal measure, and thus only a part of the community is recorded. The structures inside the mycorrhizas (hyphae, vesicles, arbuscles) are not suitable for differentiating species. So remains the molecular approach, which also has difficulties. Several "AM-specific" primers have been published, but the question is how well they work and whether all groups of Glomeromycotina are covered. Furthermore, the sequences have to be split into OTUs (**O**perational **T**axonomic **U**nits = surrogate for species)—what is the "right" threshold?!

I. Haug (✉)
Universitat Tübingen, Tübingen, Germany
e-mail: ingeborg.hang@uni-tuebingen.de

S. Setaro
Wake Forest University, Winston-Salem, USA

J. P. Suárez
Universidad Técnica Particular de Loja, Loja, Ecuador

© Springer Nature Switzerland AG 2021
R. W. Myster (ed.), *The Andean Cloud Forest*,
https://doi.org/10.1007/978-3-030-57344-7_6

As far as we know, there are only some mycorrhizal investigations in the Andean Cloud Forest. Camenzind et al. (2014) analysed the AMF community of ECSF (South Ecuador) by 454-pyrosequencing and identified 74 OTUs, among which 26 were placed in the Diversisporales and 48 in the Glomerales. Regarding their abundances, 63% of sequences represented Diversisporales. Marín et al. (2016, 2017) investigated the spore communities of AMF in *Nothofagus* and *Araucaria* forests. They proved 18 respectively 14 species, mainly members of the genus *Acaulospora*. In the Andean Yungas Forests, Geml et al. (2014) studied the community composition of all fungi in three elevation belts (400–3000 masl) based on Ion Torrent sequencing of the ITS2 rDNA from soil samples. Glomeromycota comprised only a small part of the entire soil fungal community—2.91% equalling 409 OTUs, of which 321 OTUs were Glomerales. AMF richness was negatively correlated with elevation.

In our study (Haug et al. 2019), we worked with hundreds of samples in different regions of the South Ecuadorian Andean cloud forest with the same method (nested PCR with primers AM1–AM2 [Lee et al. 2008], Sanger sequencing, threshold of 99% sequence similarity), thus providing a good database and offering the possibility to compare the communities at the different sites and elevation levels.

6.2 The Study Region

Samples were taken at four elevation belts in Southern Ecuador: 1000 masl, 2000 masl, 3000 masl and 4000 masl (Fig. 6.1, Table 6.1). The 1000 m sites are located in the Bombuscaro area in Parque Nacional Podocarpus (4°11'S, 78°96'W); the 2000 m sites in the Reserva Biológica San Francisco (RBSF; 3°58'S, 79°04'W) on the eastern slope of the Cordillera El Consuelo, Zamora-Chinchipe Province; the 3000 m sites in Cajanuma area (4°12'S, 79°17'W) in Parque Nacional Podocarpus and in the Nero area (2°95'S, 79°10'W) at Parque Nacional Cajas; and the 4000 m sites in Tutupali (3°03'S, 79°15'W), Soldados (2°98'S, 79°31'W) and the Toreadora area (2°47'S, 79°11'W) at Parque Nacional Cajas. The vegetation at 1000 masl consists of evergreen premontane rainforest, whereas the forest at 2000 masl is characterized by evergreen lower montane forest. Upper montane forest occurs at the 3000 m sites; *Polylepis* forest and grass páramo can be found at the 4000 m sites. Details are summarized in Table 6.1.

6.3 Materials and Methods (For Details See Haug et al. 2019)

At each elevation belt, we sampled three to four sites (Table 6.2). We collected arbuscular mycorrhizas directly from the soil without regarding the identity of the plant partner. Each fine-root system was placed in a separate PCR tube to ensure

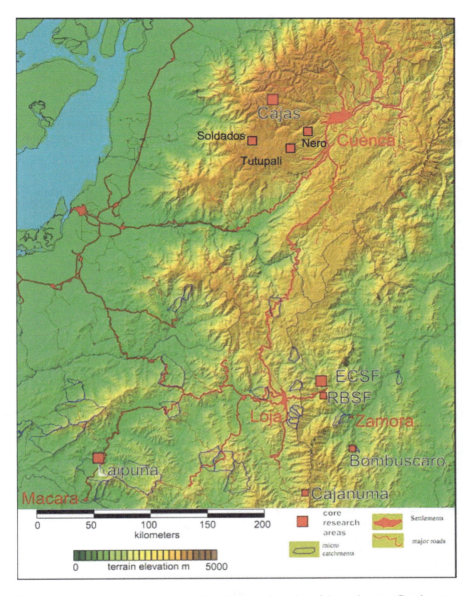

Fig. 6.1 Map of the study region in Southern Ecuador. Location of the study areas: Bombuscaro, Reserva Biologíca San Francisco (RBSF), Cajanuma, Cuenca region with Nero, Tutupali, Soldados and Cajas Toreadora. http://vhrz669.hrz.uni-marburg.de/tmf_respect/data_pre.do?citid=1745

that each mycorrhizal sample is from a single host plant. Mycorrhizas were dried at about 50°C for 24 h by placing open sample tubes on an electric dryer. After the drying step, silica gel was added and the tubes were closed for long-term storage.

In total, we successfully worked with 646 root samples: 211 at 1000 masl, 184 at 2000 masl, 128 at 3000 masl and 123 at 4000 masl (Table 6.2). Total DNA was

Table 6.1 Site characteristics for the four elevation levels. Data from (Homeier et al. 2008) (Moser et al. 2007), (Homeier et al. 2013a, b), (Martinson et al. 2013), (Crespo et al. 2011), (Pinos et al. 2017)

Elevation level	1000 m site	2000 m site	3000 m site	4000 m site
Sites	PNP—Bombuscaro	RBSF	PNP—Cajanuma	Cajas NP—Toreadora
Elevation	1050 masl	1890 masl	3060 masl	3930 masl
Vegetation	Evergreen premontane forest	Evergreen lower montane forest	Upper montane forest, shrub Páramo	Grass Páramo, *Polylepis* forest
Annual mean air temperature	19.4 °C	15.7 °C	9.4 °C	5.4 °C
Annual rainfall	c. 2230 mm	c. 1950 mm	c. 4500 mm	c. 876 mm
Soil type (FAO)	Dystric Cambisol	Stagnic Cambisol	Stagnic Histosol	Histosol, andosol
Thickness of organic layer	0 cm	10–30 cm	10–40 cm	10–40 cm
Organic layer pH-H_2O		3.9	3.9	4.2–4.8
Stand height	20–25 m, up to 40 m	18–22 m	8–10 m	

Table 6.2 Sampling subsites with number of samples taken and sequences obtained per site.

	1000 m evergreen premontane forest 1000–1140 m			2000 m evergreen lower montane forest 1900–2500 m				3000 m upper montane forest/shrub páramo 2880–3250 m					4000 m grass páramo/*Polylepis* forest 3500–4000 sm			
	B1	B2	B3	Q2	Q5	T1	T2	Cc	Cm	Cl	Clp	N	Tu	So	Cpo	Cpa
Samples analysed	103	82	26	10	63	6	105	36	21	46	17	8	20	15	32	56
Number of sequences	246	193	55	13	173	15	152	55	28	67	20	11	30	29	41	81

Abbreviations: *B1*, *B2*, *B3* Bombuscaro plot 1, 2, 3, *Q2*, *Q5* ravines, *T1*, *T2* ridges, *Cc* Cajanuma PNP Upper Montane Forest 2700 m, *Cm* Cajanuma PNP Upper Montane Forest 3000 m, *Cl* Cajanuma PNP Upper Montane Forest 2800 m, *Clp* Cajanuma PNP Shrub Páramo 3100 m, *N* Cajas NP Nero Shrub Páramo 3250 m, *Tu* Cajas NP Tutupali Grass Páramo 3500 m, *So* Cajas NP Soldados Grass Páramo 3750 m, *Cpo* Cajas NP Toreadora *Polylepis* 4000 m, *Cpa* Cajas NP Toreadora Grass Páramo 4000 m

isolated from the mycorrhizal samples with the innuPREP Plant DNA Kit (Analytik Jena, Germany). Part of the SSU region of the nuclear ribosomal rDNA repeat was amplified by PCR using a volume of 0.5 μL of the DNA template. A nested PCR approach was applied, amplifying the larger outer fragment with the primer combination NS1/NS4 and the smaller, inner fragment with the primer pair AML1/AML2. Amplified PCR products were cloned with the Invitrogen TA Cloning Kit (Life

Technologies). The final dataset consisted of 1209 Glomeromycotan sequences (Table 6.2).

All sequences were ~ 800 bp long and were aligned with MAFFT version 7 (https://mafft.cbrc.jp/alignment/server [Katoh et al. 2005], MAFFT-L-INS-i) and a distance matrix based on p-distances was created with PAUP for OTU delimitation. Operational Taxonomic Units (OTUs) were defined as surrogates for species based on 99% sequence similarity and intermediate linkage clustering with OPTSIL (Goker et al. 2010).

To explore differences in AMF community compositions among sampling sites, we carried out a non-metric multidimensional scaling (NMDS) ordination with metaMDS from the R package VEGAN (Oksanen et al. 2016).

6.4 Results

6.4.1 AMF Community in Evergreen Premontane Rain Forest at 1000 masl

Samples ($n = 211$) from the evergreen premontane forest in Bombuscaro, Podocarpus National Park, revealed 57 AMF OTUs (494 sequences). The genus *Glomus* showed the highest proportion of the overall AMF community in number of sequences (89%, Fig. 6.2a) as well as in number of OTUs (39 OTUs, 68%, Fig. 6.2d). Two *Glomus* OTUs (OTU17, OTU22) were especially abundant occurring in 39% (OTU17) and 34% (OTU22) of all samples (Fig. 6.6). The second most common genus was *Acaulospora* comprising 19.3% of all OTUs (11 total) and 8.1% of all sequences (Fig. 6.2a, d). All other genera had low OTU and low sequence numbers (Fig. 6.2a, d). Many OTUs occurred in low numbers with 78.9% of all OTUs ($n = 45$) occurring in less than 5% of all samples.

6.4.2 Ectomycorrhizas and AMF Communities in Evergreen Lower Montane Forest at 2000 masl

Three members of the Nyctaginaceae, two *Neea* species and one *Guapira* species, occurred scattered within a very species-rich evergreen lower montane forest. The three species were found to form ectomycorrhizas of very distinctive characters (Haug et al. 2005), while all other tree species examined formed arbuscular mycorrhizas. *Neea* species 1 was found to form typical ectomycorrhizas with five different fungal species, *Russula puiggarii, Lactarius* sp., two *Tomentella/Thelephora* species, and one ascomycete. *Neea* species 2 and *Guapira* species were associated with only one fungus each, a *Tomentella/Thelephora* species clustering closely together in an ITS-neighbour-joining tree (Haug et al. 2005). The long and fine rootlets of the

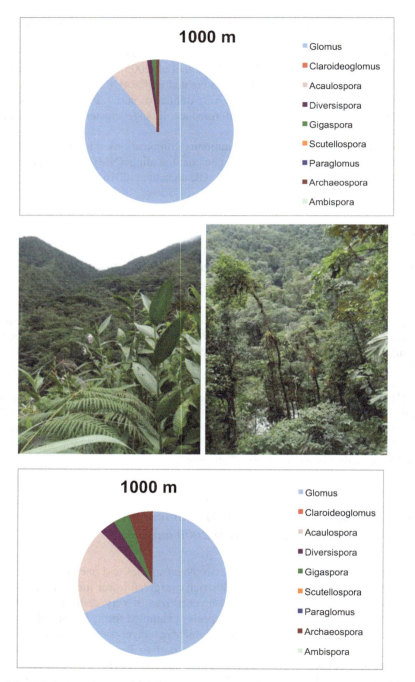

Fig. 6.2 (**a**) Relative abundance of Glomeromycotina genera in 1000 masl in number of sequences; (**b**, **c**) evergreen premontane rain forest; (**d**) relative abundance of Glomeromycotina genera in 1000 masl in number of OTUs

Guapira species showed proximally a hyphal mantle and a Hartig net, but distally intracellular fungal colonization of the epidermis and root hair development. The ectomycorrhizal segments of the long roots of *Neea* species 2 displayed a hyphal mantle and a Hartig net around alive root-hair-like outgrowths of the epidermal cells (Haug et al. 2005).

The arbuscular mycorrhizal trees at 2000 m elevation showed 66 OTUs (184 samples, 353 sequences). Seventy percent of the sequences belonged to the genus *Glomus* (Fig. 6.3a), with the most frequent *Glomus* OTUs being OTU16, OTU49 and OTU1 (Fig. 6.6). However, these OTUs occurred only in 10–13% of all samples (Fig. 6.6). *Glomus* was also the most diverse genus displaying 62% of all OTUs (Fig. 6.3d), followed by *Acaulospora* (17%) and *Archaeospora* (11%). The genera *Claroideoglomus*, *Diversispora*, *Gigaspora* and *Scutellospora* had low OTU and sequence (4%) numbers (Fig. 6.3a, d).

6.4.3 AMF Communities in Upper Montane Forest at 3000 masl

The analysis of mycorrhizal samples at 3000 m elevation revealed 37 OTUs (181 sequences in 128 samples). This number is lower than the total number of estimated OTUs occurring in the samples, as shown by the rarefaction not reaching a stable phase. Richness indices estimated 49–51 OTUs in total (Chao2, Jack1). More than half of the sequences (55%) belonged to the genus *Acaulospora* and 38% to the genus *Glomus* (Fig. 6.4a). With 16 respectively 15 OTUs, the genera *Acaulospora* and *Glomus* represented high proportions (43 + 40 = 83%) of the overall AMF community (Fig. 6.4d). All other genera showed very low OTU and sequence numbers (Fig. 6.4a, d). The most frequent OTUs were *Acaulospora* OTU87 and OTU80 occurring in 12.7% respectively 12% of all samples and *Glomus* OTU43 in 11% of samples (Fig. 6.6). As is the case for the other elevations, many OTUs were rare, with 83% of all OTUs ($n = 31$) occurring in less than 5% of all samples.

6.4.4 AMF Communities in **Polylepis** *Forest and Páramo at 4000 masl*

The samples from 4000 m sites at Cajas National Park ($n = 123$) revealed 32 OTUs (181 sequences). As is the case for the sampling at 3000 m, the rarefaction curve showed that the area is still under-sampled and richness indices estimated a total of 48–45 OTUs (Chao2, Jack1). *Acaulospora* and *Glomus* were, again, the most dominant AMF.

With 49% respectively 44%, the *Acaulospora* and *Glomus* sequences were most frequent (Fig. 6.5a). With 16 OTUs, the genus *Glomus* represented a high propor-

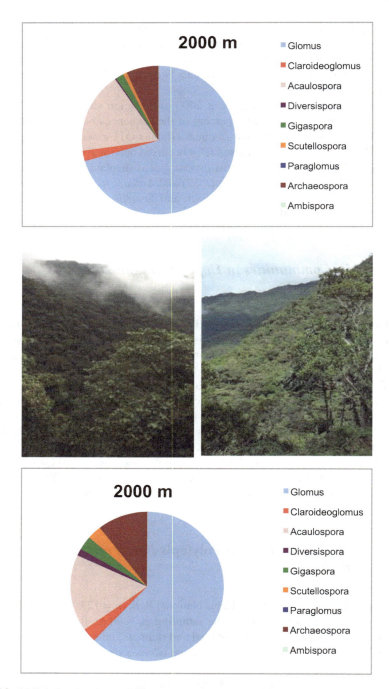

Fig. 6.3 (**a**) Relative abundance of Glomeromycotina genera in 2000 masl in number of sequences; (**b, c**) lower montane rain forest; (**d**) relative abundance of Glomeromycotina genera in 2000 masl in number of OTUs

6 Arbuscular Mycorrhizal Fungi and Ectomycorrhizas in the Andean Cloud Forest... 119

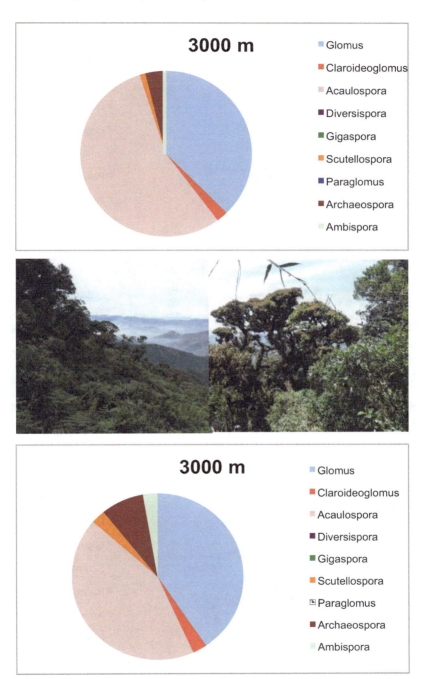

Fig. 6.4 (**a**) Relative abundance of Glomeromycotina genera in 3000 masl in number of sequences; (**b, c**) upper montane rain forest; (**d**) relative abundance of Glomeromycotina genera in 3000 masl in number of OTUs

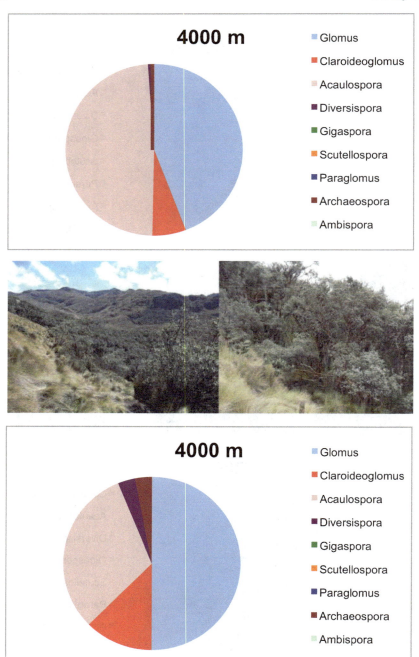

Fig. 6.5 (**a**) Relative abundance of Glomeromycotina genera in 4000 masl in number of sequences; (**b, c**) *Polylepis* forest and páramo; (**d**) relative abundance of Glomeromycotina genera in 4000 masl in number of OTUs

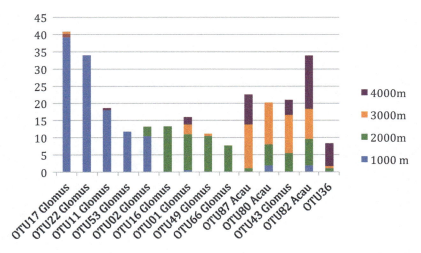

Fig. 6.6 Relative abundance of most frequent OTUs (occurrence in >10% of samples) of each altitudinal belt

tion (50%) of the overall AMF community (Fig. 6.5d). The second most common genus *Acaulospora* comprised 10 OTUs (31%, Fig. 6.5a). There was also a surprisingly high proportion of *Claroideoglomus* sequences (6,1%, Fig. 6.5a) at 4000 masl. All other genera showed low OTU and low sequence numbers (Fig. 6.5a, d). Most OTUs occurred in low numbers too, with 75% only occurring in up to 5% of the samples. The exceptions are four *Acaulospora* OTUs (82, 84, 87, 92) and four *Glomus* OTUs (23, 25, 36, 41) with occurrences ranging from 6 to 16% of all samples (Fig. 6.6).

6.4.5 AMF Community Composition Along the Elevational Gradient

Most OTUs (50%) occurred at only one altitudinal level. The highest number of these unique OTUs both in richness and frequency was in the 1000 m belt (Fig. 6.7) with a downward trend towards higher elevations (Fig. 6.7).

Three OTUs (2.6%) were found at all altitudinal levels: OTU1 (*Glomus*), OTU41 (*Glomus*) and OTU82 (*Acaulospora*); and thirteen OTUs (11%) occurred in three elevation belts.

The composition of the arbuscular mycorrhizal communities changed gradually with elevation: among the lowest (1000 masl) and highest (4000 masl) sites are only five OTUs shared and all similarity indices are low (Table 6.3, Fig. 6.7). The neighbouring altitudinal belts 1000/2000 masl and 3000/4000 masl showed a high overlap of OTUs (32%, Table 6.3, Fig. 6.7). The overlap between 2000 and 3000 m was much lower (23%), though, and was associated with a decrease in *Glomus* OTUs

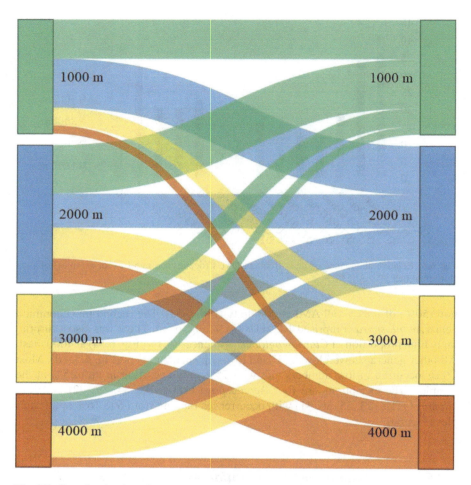

Fig. 6.7 Shared and unique OTUs of each altitudinal belt. The thickness of connector lines represents the percentage of OTUs shared

Table 6.3 Shared OTUs between elevational belts

First sample	Second sample	OTUs first sample	OTUs second sample	Shared OTUs observed
1000 m	2000 m	57	66	30 (32%)
1000 m	3000 m	57	37	11 (13%)
1000 m	4000 m	57	32	5 (6%)
2000 m	3000 m	66	37	19 (23%)
2000 m	4000 m	66	32	15 (18%)
3000 m	4000 m	37	32	19 (32%)

and an increase in *Acaulospora* OTUs on 3000 masl (Table 6.3). All similarity indices showed highest values for the neighbouring 3000 and 4000 masl communities (Table 6.4). The gradual change in OTU composition with altitude is also shown by the NMDS plot (Fig. 6.8a), which shows a high percentage of explained deviance (93%) and a mixed linearity indicated by the estimated degrees of freedom (edf) of 5.45 (edf = 1 indicates a linear relationship).

Table 6.4 Similarity indices

First Sample	Second Sample	Jaccard Classic	Sorensen Classic	Chao-Jaccard-Raw Abundance-Based	Chao-Jaccard-Est Abundance-Based	Chao-Sorensen-Raw Abundance-Based	Chao-Sorensen-Est Abundance-Based
1000 m	2000 m	0,32	0,49	0,23	0,28	0,38	0,44
1000 m	3000 m	0,13	0,23	0,16	0,45	0,27	0,63
1000 m	4000 m	0,06	0,112	0,073	0,13	0,136	0,23
2000 m	3000 m	0,226	0,369	0,31	0,534	0,473	0,696
2000 m	4000 m	0,181	0,306	0,211	0,322	0,348	0,487
3000 m	4000 m	0,38	0,551	0,55	0,74	0,71	0,85

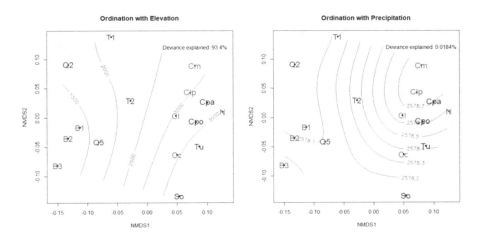

Fig. 6.8 (a, b) NMDS plot for AM fungal communities at 1000 m (blue), 2000 m (violet), 3000 m (green) and 4000 m (red) level with modelled elevation (a) and rainfall (b) mapped (pink lines) on to it. Stress value: 0.12 Abbreviations for the figures: *B1, B2, B3* Bombuscaro plot 1, 2, 3, 1000 m *Q2, Q5* ravines 2000 m, *T1* 2100 m, *T2* 2150 m ridges, *Cc* Cajanuma PNP Upper Montane Forest 2700 m, *Cm* Cajanuma PNP Upper Montane Forest 3000 m, *Cl* Cajanuma PNP Upper Montane Forest 2800 m, *Clp* Cajanuma PNP Shrub Páramo 3100 m, *N* Cajas NP Nero Shrub Páramo 3250 m, *Tu* Cajas NP Tutupali Grass Páramo 3500 m, *So* Cajas NP Soldados Grass Páramo 3750 m, *Cpo* Cajas NP Toreadora *Polylepis* 4000 m, *Cpa* Cajas NP Toreadora Grass Páramo 4000 m

6.5 Discussion

6.5.1 The Influence of Elevation on AMF Community Composition

AMF community composition is strongly influenced by elevation, which is shown by the similarity indices (Table 6.4), ordination results (Fig. 6.8) and phylogenetic community analysis. Even though few OTUs were distributed across all elevations, half of the OTUs were restricted to one height level—suggesting high community turnover among the altitudinal belts. This finding is important because it indicates that AMF communities are structured in a similar way as plant communities (Homeier et al. 2008; Homeier et al. 2013a, b).

Overall, our results showed a gradual change in AMF community composition (Figs. 6.2, 6.3, 6.4, 6.5, and 6.7). However, there was a pronounced switch from 2000 masl and below to 3000 masl and above in respect to the dominance of Glomeraceae and *Acaulospora*. Glomeraceae were dominant in lower elevations, whereas *Acaulospora* took over in higher elevations. This result was surprising because we suspected that if a shift in dominance occurred, it would be from 3000 masl to 4000 masl as this marks the switch from forest vegetation to páramo.

Many frequent OTUs from our study sites were "generalists", meaning they were widespread along the altitudinal gradient but there were just as many frequent OTUs with a narrow distribution range referred here as "specialists". Contrary to our expectations, the number and frequency of "specialist" OTUs decreased with elevation, indicating that more "generalists" and fewer "specialists" are occurring under stressful environmental conditions. This was also found by Kawahara et al. (2016) who hypothesized that AM fungi in acidic soils are specifically adapted to this soil type. Yet what they observed was that fungi in strongly acidic soils were pH generalists occurring in soils with a wide range of pH values. Thus, the conceptual framework of Gostincar et al. (2010) that fungi in extreme environments may evolve towards niche specialists unable to compete or survive in moderate environments does not seem to apply to AM fungi.

6.5.2 Species Richness Along the Elevational Gradient

Differences in observed number of OTUs may be a consequence of variable sampling intensity. OTU accumulation curves indicated that additional sampling might have resulted in the detection of more additional OTUs at 3000 and 4000 masl.

6.5.3 Abundant AMF Taxa in Our Study Sites Put in a Global Context

To put the most abundant OTUs we found in this study into a global perspective, we report here on habitat and distribution of the closest hits from the MaarjAM database. Some AMF were found to have a worldwide distribution, but some are reported here for the first time. The most abundant OTUs at the 1000 m level (OTUs 17 and 22) had no close match, neither in the MaarjAM nor in the NCBI database. One of the most widespread OTUs with global distribution and occurring in different biomes belong to the *Rhizophagus intraradices/irregularis/vesiculiferus* group (MaarjAM ID: VT113/114/115—Opik et al. 2006; Opik et al. 2010) here called OTU1. This generalist OTU was present in low numbers at all altitudinal belts but was only frequent at the 2000 m level. Another generalist and globally distributed is OTU43 (MaarjAM ID: VT191), belonging to Glomeraceae. Also found worldwide but restricted to forests is OTU16 (MaarjAM ID: VT183). It belongs to the family Glomeraceae and was only found at the 2000 m belt in our study. Another frequent 2000 m OTU was OTU49 (Glomeraceae), the type sequence for VT183 (MaarjAM ID). OTU49 might be restricted to tropical rain forests as it is so far only known from tropical rain forests in French Guiana, Gabon and Ecuador. As for *Acaulospora*, three OTUs were frequent in our study area: OTU80 (Maarjam ID: VT12), occurring from 1000 m to 3000 m but with main abundance in 3000 m, is otherwise only known from South America (Argentina and French Guiana); OTU 82 (Maarjam ID: VT14), a generalist that also occurs in Europe and Asia; and OTU87 (Maarjam ID: VT30), in our study only at 3000 m but with worldwide distribution and in many biomes.

The dominance of Glomeraceae at 1000 and 2000 masl is in accordance with many other studies using spore-, Sanger-sequence- or NGS-based sampling techniques from different ecosystems (e.g. Gai et al. 2006, Camenzind et al. 2014, Leal et al. 2013, da Silva et al. 2015, Rodriguez-Echeverria et al. 2017, Geoffroy et al. 2017). For *Acaulospora*, however, the literature is controversial. Dominance of Acaulosporaceae in higher elevations, as our results show, is supported by Egan et al. (2017), who observed an increase in the number of *Acaulospora* species along a high elevation gradient in Montana (USA) and in three studies who found the same pattern in the Himalaya (Yang et al. 2016; Li et al. 2014; Liu et al. 2015). Also, two unknown *Acaulospora* species were the dominant colonizers of Andean potatoes growing between 2658 and 4075 masl (Senes-Guerrero and Schussler 2016). Other studies in the Himalaya did not find *Acaulospora* to be dominant, but Glomeraceae instead (Li et al. 2015; Yang et al. 2016; Kotilinek et al. 2017). The reason behind the frequency patterns of Glomeraceae and Acaulosporaceae is not clear and remains to be studied.

6.5.4 Rare OTUs in AMF Communities and Its Potential Role in Ecosystem Functioning

Like previous studies of AM fungal community structure (Rosendahl 2008; Dumbrell et al. 2010), we found a high proportion of rare OTUs at all altitudes: 79–85% of OTUs occurred in less than 5% of the samples. However, abundant taxa representing on average 40% of total abundance within the community as shown in previous studies (Dumbrell et al. 2010) were only present at the 1000 m level. At higher altitudes, the most common OTUs were detected in only 13–15% of the samples, which is not very frequent. This mirrors the situation for plants in tropical forests (Wright 2002). The processes behind these patterns are not fully understood, not even for plants (Prada and Stevenson 2016). Niche differentiation is considered to be an important factor in the structuring of communities, for plants (Prada and Stevenson 2016; Wright 2002) and AM fungi (Zobel and Opik 2014), but stochastic processes also play a role (Lekberg and Waller 2016; Encinas-Viso et al. 2016). The study sites we investigated had a broad range of edaphic variability at each altitudinal belt, which could provide many microhabitats for AMF to specialize in. Examples are ridges vs. ravine habitats at the 2000 m belt and grassland páramo vs. *Polylepis* forest at 4000 masl.

The importance of rare plant species and their distributions across ecosystems is discussed in several studies (e.g. Gaston 2012; Mi et al. 2012). Determining the ecological similarity of rare and common species is not easy even in plants, particularly when a large number of species are involved. We can assume, though, that the more the species and the higher the phylogenetic diversity, the more traits are available in a community. The multitude of rare fungi in our study with a broad phylogenetic level at each elevation thus offers a wide range of traits with potential benefits to their plant partners. Therefore, rare species might be important for maintaining ecosystem functioning in a changing environment (Bachelot et al. 2015). We can see evidence for this in our data as some OTUs (e.g. OTU1, OTU43, OTU80, OTU82, OTU87) are rare at a certain altitude but become dominant in another altitude. The ability to maintain many rare OTUs in combination with low specialization might be a reason why AM fungi have been one of the most successful organisms on the earth, being around since 450 million years and occurring in all terrestrial habitats.

Acknowledgements We thank Jutta Bloschies for laboratory work. We appreciate the German Research Foundation (DFG) for financing the study (Research Units 402 and 816, PAK825), Naturaleza y Cultura Internacional (Loja, San Diego) for support, our Ecuadorian partner Universidad Técnica Particular de Loja (UTPL) for outstanding cooperation, the Ecuadorian Ministerio del Ambiente for granting research permits, Patrick Hildebrandt and Carlos Quiroz for the possibility to take samples in Nero, Soldados and Tutupali, and the staff of the Estación Científica San Francisco—especially Felix Matt and Jörg Zeilinger—for their great assistance.

References

Bachelot B, Kobe RK, Vriesendorp C (2015) Negative density-dependent mortality varies over time in a wet tropical forest, advantaging rare species, common species, or no species. Oecologia 179(3):853–861

Camenzind T, Hempel S, Homeier J, Horn S, Velescu A, Wilcke W, Rillig MC (2014) Nitrogen and phosphorus additions impact arbuscular mycorrhizal abundance and molecular diversity in a tropical montane forest. Glob Chang Biol 20(12):3646–3659

Crespo PJ, Feyen J, Buytaert W, Bucker A, Breuer L, Frede HG, Ramirez M (2011) Identifying controls of the rainfall-runoff response of small catchments in the tropical Andes (Ecuador). J Hydrol 407(1–4):164–174

da Silva DKA, Coutinho FP, Escobar IEC, de Souza RG, Oehl F, Silva GA, Cavalcante UMT, Maia LC (2015) The community of arbuscular mycorrhizal fungi in natural and revegetated coastal areas (Atlantic Forest) in northeastern Brazil. Biodivers Conserv 24(9):2213–2226

Dumbrell AJ, Nelson M, Helgason T, Dytham C, Fitter AH (2010) Idiosyncrasy and overdominance in the structure of natural communities of arbuscular mycorrhizal fungi: is there a role for stochastic processes? J Ecol 98(2):419–428

Egan CP, Callaway RM, Hart MM, Pither J, Klironomos J (2017) Phylogenetic structure of arbuscular mycorrhizal fungal communities along an elevation gradient. Mycorrhiza 27(3):273–282

Encinas-Viso F, Alonso D, Klironomos JN, Etienne RS, Chang ER (2016) Plant-mycorrhizal fungus co-occurrence network lacks substantial structure. Oikos 125(4):457–467

Frank B (1885) Über die auf Wurzelsymbiose beruhende Ernährung gewisser Bäume durch unterirdische Pilze. Ber. Deutsch. Bot Ges 3:128–145

Gai JP, Christie P, Feng G, Li XL (2006) Twenty years of research on community composition and species distribution of arbuscular mycorrhizal fungi in China: a review. Mycorrhiza 16(4):229–239

Gaston KJ (2012) ECOLOGY the importance of being rare. Nature 487(7405):46–47

Geml J, Pastor L, Fernandez S, Pacheco TA, Semenova AG, Becerra CY, Wicaksono W, Nouhra ER (2014) Large-scale fungal diversity assessment in the Andean Yungas forests reveals strong community turnover among forest types along an altitudinal gradient. Mol Ecol 23(10):2452–2472

Geoffroy A, Sanguin H, Galiana A, Ba A (2017) Molecular Characterization of Arbuscular Mycorrhizal Fungi in an Agroforestry System Reveals the Predominance of Funneliformis spp Associated with Colocasia esculenta and Pterocarpus officinalis Adult Trees and Seedlings. Front Microbiol 8:1426

Goker M, Grimm GW, Auch AF, Aurahs R, Kucera M (2010) A clustering optimization strategy for molecular taxonomy applied to planktonic foraminifera SSU rDNA. Evol Bioinform Online 6:97–112

Gostincar C, Grube M, de Hoog S, Zalar P, Gunde-Cimerman N (2010) Extremotolerance in fungi: evolution on the edge. FEMS Microbiol Ecol 71(1):2–11

Haug I, Setaro S, Suarez JP (2019) Species composition of arbuscular mycorrhizal communities changes with elevation in the Andes of South Ecuador. PLoS One 14(8):e0221091

Haug I, Weiss M, Homeier J, Oberwinkler F, Kottke I (2005) Russulaceae and Thelephoraceae form ectomycorrhizas with members of the Nyctaginaceae (Caryophyllales) in the tropical mountain rain forest of southern Ecuador. New Phytol 165(3):923–936

Homeier JL, Bräuning A, Cumbicus NL, Hertel D, Matrinson GO, Spannl S, Veldkamp E (2013a) Effects of nutrient addition on the productivity of montane forests and implications for the carbon cycle. In: Bendix JB, Bräuning E, Makeschin A, Mosandl F, Scheu R, Wilcke W (eds) Ecosystem services, biodiversity and environmental change in a tropical mountain ecosystem of South Ecuador. Springer Verlag, Heidelberg, pp 315–329

Homeier JW, Gawlik FA, Peters J, Diertl T, Richter M (2013b) Plant diversity and its relevance for the provision of ecosystem services. In: Bendix JB, Bräuning E, Makeschin A, Mosandl

F, Scheu R, Wilcke W (eds) Ecosystem services, biodiversity and environmental change in a tropical mountain ecosystem of South Ecuador. Springer Verlag, Heidelberg, pp 93–106

Homeier JW, Gradstein FA, Breckle SR, Richter M (2008) Potential vegetation and floristic composition of Andean forests in South Ecuador, with a focus on the RBSF. In: Beck EB, Kottke J, Makeschin I, Mosandl F (eds) Gradients in a tropical mountain ecosystem of Ecuador. Springer Verlag, Berlin, pp 87–100

Katoh K, Kuma K, Toh H, Miyata T (2005) MAFFT version 5: improvement in accuracy of multiple sequence alignment. Nucleic Acids Res 33(2):511–518

Kawahara A, An GH, Miyakawa S, Sonoda J, Ezawa T (2016) Nestedness in arbuscular mycorrhizal fungal communities along soil pH gradients in early primary succession: acid-tolerant Fungi are pH generalists. PLoS One 11(10):e0165035

Kotilinek M, Hiiesalu I, Kosnar J, Smilauerov M, Smilauer P, Altman J, Dvorsky M, Kopecky M, Dolezal J (2017) Fungal root symbionts of high-altitude vascular plants in the Himalayas. Sci Rep 7:6562

Leal PL, Siqueira JO, Sturmer SL (2013) Switch of tropical Amazon forest to pasture affects taxonomic composition but not species abundance and diversity of arbuscular mycorrhizal fungal community. Appl Soil Ecol 71:72–80

Lee J, Lee S, Young JPW (2008) Improved PCR primers for the detection and identification of arbuscular mycorrhizal fungi. FEMS Microbiol Ecol 65(2):339–349

Lekberg Y, Waller LP (2016) What drives differences in arbuscular mycorrhizal fungal communities among plant species? Fungal Ecol 24:135–138

Li XL, Gai JP, Cai XB, Li XL, Christie P, Zhang FS, Zhang JL (2014) Molecular diversity of arbuscular mycorrhizal fungi associated with two co-occurring perennial plant species on a Tibetan altitudinal gradient. Mycorrhiza 24(2):95–107

Li XL, Zhang JL, Gai JP, Cai XB, Christie P, Li XL (2015) Contribution of arbuscular mycorrhizal fungi of sedges to soil aggregation along an altitudinal alpine grassland gradient on the Tibetan plateau. Environ Microbiol 17(8):2841–2857

Liu L, Hart MM, Zhang JL, Cai XB, Gai JP, Christie P, Li XL, Klironomos JN (2015) Altitudinal distribution patterns of AM fungal assemblages in a Tibetan alpine grassland. FEMS Microbiol Ecol 91(7):fiv078

Marín C, Aguilera P, Cornejo P, Godoy R, Oehl F, Palfner G, Boy J (2016) Arbuscular mycorrhizal assemblages along contrastingAndean forests of southern Chile. J Soil Sci Plant Nutr 16(4):916–929

Marín, C., P. Aguilera, , F. Oehl and R. Godoy (2017). Factors affecting arbuscular mycorrhizal fungi of Chilean temperate rainforests. J Soil Sci Plant Nutr 17 (4): 966–984

Martinson GO, Corre MD, Veldkamp E (2013) Responses of nitrous oxide fluxes and soil nitrogen cycling to nutrient additions in montane forests along an elevation gradient in southern Ecuador. Biogeochemistry 112(1–3):625–636

Mi XC, Swenson NG, Valencia R, Kress WJ, Erickson DL, Perez AJ, Ren HB, Su S, Gunatilleke N, Gunatilleke S, Hao ZQ, Ye WH, Cao M, Suresh HS, Dattaraja HS, Sukumar MKP (2012) The contribution of rare species to community phylogenetic diversity across a global network of Forest plots. Am Nat 180(1):E17–E30

Moser G, Hertel D, Leuschner C (2007) Altitudinal change in LAI and stand leaf biomass in tropical montane forests: a transect shady in Ecuador and a pan-tropical meta-analysis. Ecosystems 10(6):924–935

Oksanen J, Blanchet FG, Friendly M, Kindt R, Legendre P, McGlinn D (2016) Vegan: community Ecology package. R package version 2:4–1

Opik M, Moora M, Liira J, Zobel M (2006) Composition of root-colonizing arbuscular mycorrhizal fungal communities in different ecosystems around the globe. J Ecol 94(4):778–790

Opik M, Vanatoa A, Vanatoa E, Moora M, Davison J, Kalwij JM, Reier U, Zobel M (2010) The online database MaarjAM reveals global and ecosystemic distribution patterns in arbuscular mycorrhizal fungi (Glomeromycota). New Phytol 188(1):223–241

Pinos J, Studholme A, Carabajo A, Gracia C (2017) Leaf Litterfall and decomposition of Polylepis reticulata in the Treeline of the Ecuadorian Andes. Mt Res Dev 37(1):87–96

Prada CM, Stevenson PR (2016) Plant composition associated with environmental gradients in tropical montane forests (Cueva de Los guacharos National Park, Huila, Colombia). Biotropica 48(5):568–576

Rodriguez-Echeverria S, Teixeira M, Correia S, Timoteo R, Heleno M, Opik M, Moora M (2017) Arbuscular mycorrhizal fungi communities from tropical Africa reveal strong ecological structure. New Phytol 214(1):487–487

Rosendahl S (2008) Communities, populations and individuals of arbuscular mycorrhizal fungi. New Phytol 178(2):253–266

Senes-Guerrero C, Schussler A (2016) A conserved arbuscular mycorrhizal fungal core-species community colonizes potato roots in the Andes. Fungal Divers 77(1):317–333

Wright SJ (2002) Plant diversity in tropical forests: a review of mechanisms of species coexistence. Oecologia 130(1):1–14

Yang W, Zheng Y, Gao C, Duan JC, Wang SP, Guo LD (2016) Arbuscular mycorrhizal fungal community composition affected by original elevation rather than translocation along an altitudinal gradient on the Qinghai-Tibet Plateau. Sci Rep 6:36606

Zobel M, Opik M (2014) Plant and arbuscular mycorrhizal fungal (AMF) communities - which drives which? J Veg Sci 25(5):1133–1140

Chapter 7
Nesting Ecology of the Tucuman Amazon (*Amazona tucumana*) in the Cloud Forest of Northwestern Argentina

Luis Rivera and Natalia Politi

7.1 The Tucuman Amazon

Parrots (Order Psittaciformes) are one of the largest groups of birds, with 398 species distributed in tropical and subtropical areas of the world and are among the most threatened avian orders with 30% of the species endangered at global level from which 15% are Neotropical (BirdLife International 2018). Parrots of the genus *Amazona* are 27–31 species broadly distributed in Central and South America (México to Argentina) and the Caribbean that inhabit from lowlands to highlands, and from dry to wet habitats (Juniper and Parr 1998). All parrot species of this genus were heavily traded in the international pet market, are currently included in the CITES appendix I and II, and 18 species are categorized as threatened (Russello and Amato 2004, BirdLife Internationa 2018).

The Tucuman Amazon (*A. tucumana*) is one of the four *Amazona* parrot species from Argentina (Narosky and Yzurieta 2003), which was shown to be sister species of the Spectacled Amazon (*Amazona pretrei*) (Russello and Amato 2004; Rocha et al. 2014). The group formed by Tucuman Amazon, Spectacled Amazon, and Vinaceous-breasted Amazon probably is a basal group for all the other species from the genus *Amazona* (Russello and Amato 2004). Recent genetic studies have allowed assessing allopatric models to better explain the diversification process of Spectacled Amazon and Tucuman Amazon, to determine genetic variability, effective population size, and divergence time of lineage separation (Rocha et al. 2014). Haplotype diversity in Tucuman Amazon (DH = 0.41) is smaller than in Spectacled Amazon (DH = 0.68), the effective population size of females for Tucuman Amazon was approximately 94,000 individuals, and the divergence time of lineage separation was estimated to be approximately 1.3 million years ago, which corresponds to the lower Pleistocene (Rocha et al. 2014). The genetic information suggest allopatry by

---

L. Rivera (✉) · N. Politi
Insituto de Ecorregiones Andinas CONICET/Universidad Nacional de Jujuy,
S.S. de Jujuy, Jujuy, Argentina

© Springer Nature Switzerland AG 2021
R. W. Myster (ed.), *The Andean Cloud Forest*,
https://doi.org/10.1007/978-3-030-57344-7_7

the dispersion/founder effect model (peripheral isolation) as the more likely process to explain the diversification of these two Amazon species (Rocha et al. 2014).

As most Amazon species Tucuman Amazon sexes are alike (Forshaw 1977). Tucuman Amazon total body length is 31 cm, weights 280 g, and has a short tail. The head of mature individuals of Tucuman Amazon shows a front patch of red feathers, and juveniles show yellow feathers around this patch. Also the primary covers in the wings form a patch of red feathers (speculum) visible in flight. The tip of the wing feathers is purple blue and the body feathers are strongly emarginated with dark grey (Fig. 7.1).

Tucuman Amazon was described by Cabanis in 1885 since then, little information has been obtained about its natural history and ecology and anecdotical mentions about its biology are known (Hoy 1968; Orfila 1938; Wetmore 1926). Diet of the Tucuman Amazon includes seeds of *Alnus acuminata*, *Podocarpus parlatorei*, fruits of *Cedrela lilloi*, and flowers of *Erythrina falcata* (Fjeldså and Krabbe 1990; Juniper and Parr 1998; Low 2005). In autumn and winter the species can feed on groves and crops (Low 2005) when it descend to lower elevations even visiting fruiting trees in cities (Fjeldså and Krabbe 1990, Low 2005, Juniper and Parr 1998).

Fig. 7.1 Tucuman Amazon and plumage characteristics

7 Nesting Ecology of the Tucuman Amazon (*Amazona tucumana*) in the Cloud Forest... 133

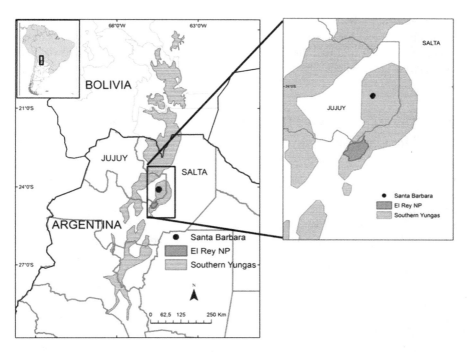

Fig. 7.2 Distribution of the Southern Yungas or Tucumano-Boliviano forest. Study sites include El Rey National Park and Santa Barbara

The Tucuman Amazon is a threatened species (BirdLife International 2018), with a small geographic range restricted to the narrow strip of montane forest on the eastern slopes of the Andes (the Yungas) from south-eastern Bolivia to northwestern Argentina (Fjeldså and Krabbe 1990) (Fig. 7.2). The Yungas is a biogeographic province belonging to the Amazonic domain that spread along the eastern slopes of the Andes from Venezuela to northwestern Argentina encompassing an elevation range between 400 and 3500 m a.s.l. (Cabrera and Willink 1980). Climate is highly seasonal with rainfall concentrated (75–80%) during the summer (i.e., November to March) (Mendoza 2005). The southernmost limit of Neotropical montane forests is known as Southern Yungas or Tucumano-Boliviano forest (Fig. 7.2, Tortorelli 1956, Hueck 1978, Cabrera 1994). In Bolivia, the Southern Yungas goes from 18° S through Tarija, Chuquisaca, and Santa Cruz de la Sierra departments and in Argentina it extends to 29° S, through Catamarca, Tucumán, Salta, and Jujuy provinces. Southern Yungas are distributed discontinuously by 1200 km forming isolated patches of 50 km wide covering an area of approx. four million hectares in Argentina (Brown 1995), and three million hectares in Bolivia (Ibisch and Mérida 2003). Along the Southern Yungas there is a latitudinal gradient (14° S–29° S) with a marked loss of plant and animal species richness as latitude increases (Brown et al. 2001). In Argentina, the Southern Yungas represents 2% of the terrestrial land but holds the higher level of endemism of flora and fauna of the country (Kappelle and Brown 2001). In the Southern Yungas the breeding habitat of Tucuman Amazon is the cloud forest found between 1500 and 2200 m asl that has an annual rainfall averages of 1300 mm and annual temperature averages of 11.7 °C (Brown et al. 2001; Arias and Bianchi 1996).

7.2 The Reproductive Habitat of Tucuman Amazon: The Cloud Forest

Of the eight parrot species that inhabit the Southern Yungas only the Tucuman Amazon and the Red-mitred parakeet are sympatric in the cloud forest. The remaining six species occupy other forest types along the elevation gradient of the mountain ranges (Politi and Rivera 2005). The forest structure in one old-growth cloud forest in El Rey National Park (Rivera 2011) is characterized with a density of 406 ± 136 trees/ha >10 cm DBH and a basal area of 45.2 ± 21.4 m^2/ha. The size distribution for trees >10 cm DBH has an inverse J form typical of multispecies stands with larger numbers of small stems and a diminishing density of larger ones as DBH increase (Fig. 7.3). Species richness of trees >10 cm DBH is of 19 species (Table 7.1). Twelve tree species were dominant with 97.3% of the individuals >10 cm DBH and 7 tree species showed very low densities. Dominant tree species are *Blepharocalyx salicifolius, Podocarpus parlatorei, Cinnamomum porphyria,*

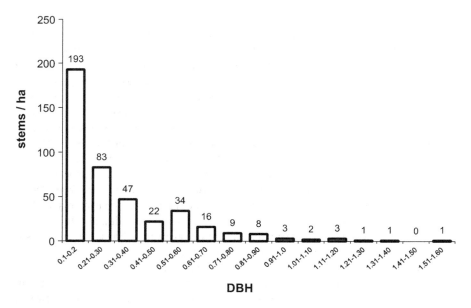

Fig. 7.3 Forest structure in DBH size classes of the stems >10 cm for all tree species. Number above bar is the total number of stems

Table 7.1 Tree species composition of live trees >10 cm DBH and snags in El Rey National Park, Argentina. Tree density, DBH, height, and basal area expressed as mean ± SE

Tree species	Frequency (%)	Density (N/ha)	DBH (cm)	Height (m)	Basal area (m²/ha)
Myrcianthes pseudomato	21.7	90.0 ± 15.6	27.9 ± 2.6	13.5 ± 0.4	3.42 ± 0.6
Blepharocalyx salicifolius	16.6	69.0 ± 12.1	38.0 ± 3.3	19.8 ± 1.0	12.44 ± 3.7
Allophylus edulis	14.0	59.0 ± 12.9	28.7 ± 3.5	12.6 ± 0.4	1.62 ± 0.4
Podocarpus parlatorei	8.9	37.0 ± 15.1	42.8 ± 4.8	17.4 ± 0.6	7.0 ± 2.4
Cinnamomum porphyria	6.5	27.0 ± 7	28.1 ± 3.6	21.7 ± 0.5	4.56 ± 1.3
Myrrhinium atropurpureum	6.3	26.0 ± 7.2	19.1 ± 1.1	12.2 ± 1.1	0.56 ± 0.2
Snags[1]	6.0	25.0 ± 5.9	33.1 ± 5.6	–	3.08 ± 1.1
Cedrela lilloi	4.3	18.0 ± 4.8	45.5 ± 4.9	24.4 ± 2.2	8.18 ± 2.8
Prunus tucumanensis	4.1	17.0 ± 6.2	26.0 ± 2.3	12.4 ± 0.9	1.0 ± 0.4
Juglans australis	3.4	14.0 ± 2.6	42.4 ± 4.9	23.0 ± 1.8	2.34 ± 0.6
Sambucus peruviana	1.9	8.0 ± 3.9	17.0 ± 1.4	–	0.18 ± 0.1
Ilex Argentina	1.7	7.0 ± 3	21.7 ± 1.8	14.2 ± 2.1	0.26 ± 0.1
Myrcianthes sp.	1.7	7.0 ± 2.6	15.9 ± 2.0	12.2 ± 1.1	0.14 ± 0.1

[1] Snags represent standing dead trees of different tree species, given that after trees die it is difficult to identify.

Cedrela lilloi, Juglans australis, Myrcianthes pseudomato, Allophylus edulis, Myrrhinium atropurpureum, Prunus tucumanensis, Sambucus peruviana, Ilex argentina, and Myrcianthes callicoma. Mean height of trees >10 cm DBH was 16.5 ± 5.9 m (Range = 6–34 m). The forest shows three well defined stratum with a continuous canopy formed by trees between 10 and 20 m height, some emergent trees higher than 20 m, and a lower stratum of smaller trees between 0 and 10 m height forming the subcanopy, and an understory of saplings, ferns, shrubs, and herbs (Rivera 2011).

Analysis of availability of Tucuman Amazon habitat at the eco-regional level using distribution models indicates that conditions for supporting Tucuman Amazon habitat were distributed across the entire north–south gradient of the Southern Yungas, but less than half of the Southern Yungas (~46,000 km², 42%) was predicted to provide suitable non-breeding habitat, with slightly more occurring in Bolivia (~26,000 km²) than in Argentina (~20,000 km²) (Pidgeon et al. 2015). Potential breeding habitat is more limited than non-breeding habitat since conditions that could support breeding habitat were found in only ~21,000 km², or 19%, of the Southern Yungas with three times as much breeding habitat found in Argentina (~15,400 km²) than in Bolivia (~5600 km²) (Pidgeon et al. 2015). In Bolivia, potential breeding habitat existed only in very narrow belts and disjunct patches along ridgetops, whereas in Argentina potential breeding habitat patches were larger, with a more regular shape (Pidgeon et al. 2015). Yet only 15% (3134 km²) of the breeding habitat is currently within protected areas, and only ~6%, all in Argentina, is strictly protected.

From the conservation side the main factors affecting the Tucuman Amazon are natural habitat destruction and fragmentation, as well as the illegal trafficking of nestlings and adults (Martinez and Prestes 2008; Rivera et al. 2007). Tucuman Amazon has a small population size, with population estimates have identified approximately 8000 individuals, 80% of which are in Argentina in the northern and central distribution (Rivera et al. 2007; Rivera et al. 2010). The low genetic variability observed in this species (DH = 0.41) may be mainly the result of its evolutionary history, due to the founding event in the early formation of this species (Rocha et al. 2014). Furthermore, perhaps in recent centuries, it could have also been influenced by the different negative impacts of human actions, making the species more genetically vulnerable. Thus, conservation efforts for the Tucuman Amazon should prioritize the northern and central regions of Southern Yungas in Argentina, where 80% of the remaining Tucuman Amazon population occurs, with high priority given to the creation of public and private reserves (Rivera et al. 2007). Furthermore, this species may require actions aimed at increasing the population size, such as eliminating nest poaching and retaining nest and food trees in forestry operations (Rivera et al. 2012; Rivera et al. 2013). Anecdotal evidence of Tucuman Amazon kept as pets in villages in Bolivia suggests that local customs and attitudes toward parrots are a challenge to the conservation of this species (Tella et al. 2013). Finally, the low genetic variability and genetic distinctiveness in Tucuman Amazon can be seen as very valuable information given the categorization of this species on the IUCN Red List as Vulnerable (BirdLife International 2018).

Information on habitat requirements allows predictions to be made on the ability of Tucuman Amazon to adapt to disturbed habitats (Saunders et al. 1982) and to develop effective strategies for conservation and management of threatened species (Renton 2000). It is necessary to conduct studies on mature or old-growth forests (sensu Hunter Jr. and White 1997) that set a reference for future comparisons against habitat modification. Detailed knowledge of breeding biology is necessary for understanding variation in avian reproductive strategies, because it provides critical natural history data that are useful for generating new hypotheses and testing old ones (Auer et al. 2007). The lack of information on the natural history, ecology, and demography of many parrot species precludes an assessment of the mechanisms that regulate population dynamics (Koenig 2001). Prior to this study there were no specific studies on the reproductive biology of Tucuman Amazon—there was one record of a nest, found in Chuquisaca, Bolivia, with a female incubating four eggs in January (Bond and Meyer de Schauensee 1943); a note that Tucuman Amazon nests in large trunks of *Alnus* or *Podocarpus* trees between January and March in forests located at an elevation of 2600 m (Juniper and Parr 1998); and a comment that Tucuman Amazon places its nests at elevations between 900 and 2100 m (Low 2005). In captivity, Tucuman Amazon brood mean size was 3.4 eggs with a development period of the chicks lasting between 60 and 67 days (Low 2005).

7.3 Nesting Habitat Requirements of Tucuman Amazon

To characterize the nesting habitat of Tucuman Amazon we assessed cavity availability, reuse, and spatial pattern of nests in an old-growth forest in the central sector of the Southern Yungas of northwestern Argentina, on the eastern slopes of the Sierras Subandinas Centrales or Sistema de Santa Bárbara—a mountain range ~ 100 km long, between the Cordillera Oriental to the west and the Chaco plain to the east. Within the central sector of the Southern Yungas, we focused on two areas where we knew of active nests: (1) El Rey National Park, Salta Province (24° 43'S, 64° 38'W, 44,000 ha) and (2) Portal de Piedra Private Reserve, Jujuy Province (24° 05'S, 64° 26'W, 400 ha). Both areas are strictly protected and controlled, no people live within the reserves and, to the best of our knowledge, there is no poaching. The study was conducted within accessible areas only. These comprised ~170 ha in Portal de Piedra Private Reserve and 45 ha in El Rey National Park. Elevation of these areas is between 1450 and 2100 m above sea level.

We carried out fieldwork from 2004 to 2009. Nest searches of Tucuman Amazon were conducted daily during egg-laying and incubation periods (December to mid-January). We found nests by following males to the nest area and locating the cavity when the female left the nest to be fed by the male (González Elizondo 1998). Nest-cavity characteristics were determined by climbing (Perry 1978) and measuring cavity dimensions (Fig. 7.4). We recorded the following nest characteristics: (1) height from the ground to the cavity entrance; (2) size of cavity entrance (horizontal and vertical); (3) internal diameter at the cavity floor; (4) internal cavity depth from

Fig. 7.4 Nest monitoring of Tucuman Amazon and nest entrance in *Podocarpus parlatorei* tree

cavity entrance to the floor; (5) trunk or branch diameter at cavity entrance; (6) tree diameter at the cavity floor; (7) tree diameter at breast height (DBH); (8) tree height; (9) tree species; (10) cavity origin (excavated or decayed); (11) cavity location (tree trunk, primary branch, secondary branch, or third branch); (12) tree status (alive or dead); and (13) tree location (latitude and longitude). Our characterization of nesting requirements is based on 44 active nests found in El Rey National Park. To assess tree-cavity availability we conducted sampling during the non-breeding season (April–August 2007 and 2008) when many trees are leafless. We used distance sampling methodology to estimate the density of suitable cavities. We performed 20 variable-width, random direction, 300-m long transects that were at least 150 m apart. We measured the perpendicular distance from the central line of the transect to each detected cavity. We only considered a cavity to be suitable if it had a hollow chamber surrounded by sound wood (not collapsing wood), accessed by entrance holes with a floor to support an incubation chamber and a roof to provide overhead protection, a minimum diameter entrance of 5 cm, an internal diameter of at least 15 cm (minimum cavity dimensions suggested for *Amazona* species of similar body-size to Tucuman Amazon; Snyder et al. 1987, Enkerlin-Hoeflich 1995), a minimum cavity height from the ground of 2 m, cavity depth from 0 to 200 cm, and a tree DBH of 30 cm (minimum dimensions observed for Tucuman Amazon in another site; Rivera 2011). Therefore, the minimum characteristics used to determine a suitable cavity were in the range of the cavities used for nesting. We used a tree-peeper (Richardson et al. 1999) to estimate or measure the following cavity and tree characteristics: (1) height from the ground to the cavity entrance using the graduated metric scale in the telescopic rod of the tree-peeper; (2) cavity entrance diameters (horizontal and vertical); (3) internal diameter at the cavity floor; (4) internal cavity depth from cavity entrance to the floor. Cavity entrance bearing was measured with a compass, tree DBH was measured with metric tape and tree height with a hypsometer. Due to tree-peeper limitations we only inspected suitable cavities below 15 m (Richardson et al. 1999).

We used Manly's selection index to compare use of cavities as nest-sites in different tree species with the availability of cavities in those tree species (Krebs 1999; Manly et al. 2002). We calculated a selection coefficient and the 95% confidence interval for the categorical nest-site variable (tree species). Coefficients greater than 1.0 indicated preference, while values less than 1.0 indicated avoidance (Krebs 1999; Manly et al. 2002; Aitken and Martin 2004). Selection coefficients were tested for significance using the log-likelihood ratio (G-test, Manly et al. 2002). Frequencies of nest cavities in different categories (tree species, origin, and cavity location) were compared with a χ^2 test. We determined the number of cavities reused by Tucuman Amazon for nesting over several breeding season. We define reuse as those cases where the same cavity was used in more than 1 year (Berkunsky and Reboreda 2009) and a cavity was considered to be used if it contained eggs or chicks.

Cavity density were determined following line transect guidelines and modelled using the software Distance 5.0 (Buckland et al. 2001; Thomas et al. 2006). The model with the lowest Akaike's Information Criterion (AIC) was selected (Burnham and Anderson 2002). The adequacy of the selected model for the perpendicular distances was assessed using a Kolmogorov–Smirnov test (Buckland et al. 2001). We determined the average nest density by calculating the mean of the number of nests found in 45 ha during the four breeding seasons. We used the Spatial Analyst tool of ArcGIS to determine distances to evaluate spacing among all simultaneously active nests and using the locations of all trees used as nest-sites over the four-year study (Salinas-Melgoza et al. 2009). Each nest-tree location was considered only once for the analysis regardless of how many times the tree was reused as a nest-site. In addition, for each nest-tree used by parrots in any year we calculated the distance to the nearest neighbouring tree that had been used as a nest-site in any year. We compared the nearest neighbour distances for active nests among breeding seasons with a Kruskal–Wallis test. To determine whether the spacing of breeding pairs differed from the distribution of all nest-trees we compared distances among active breeding pairs with distances among nest-tree for all years combined, using a Mann–Whitney U test. Using a paired Wilcoxon test, we further evaluated the influence of conspecifics on the spacing of parrot nests to compare the distance to the nearest active nest vs. the distance to the nearest potential unoccupied nest-tree for each parrot nest active in the 2008–2009 breeding season. We restricted this analysis to the 2008–2009 datasets, which had the most complete record of potential nest-trees, to avoid overduplication of distance values between years (Salinas-Melgoza et al. 2009). Distance values obtained previously among all nests were used to assess the spatial pattern of nest-bearing trees and active nests (Salinas-Melgoza et al. 2009) with the Average Nearest Neighbor Distance tool from ArcGIS (Mitchell 2005). All the values are expressed as mean standard deviation (SD) unless otherwise specified. We set the significance level of statistical tests at $P = 0.05$.

We recorded 44 Tucuman Amazon nesting attempts in 37 nest-trees, 30 during incubation, and seven during brooding. Most Tucuman Amazon nests occurred in live trees (95%) of six species, and only 5% were in nest cavities in snags. There was a significant difference in the frequency of tree species used for nesting ($\chi^2 = 27.6$, $P < 0.001$), with most nest cavities in *Blepharocalyx salicifolius* (59.5%, 22 out of 37), followed by Juglans australis (13.5%, 5), Podocarpus parlatorei (8.5%, 3), *Cinnamomum porphyria* (5.4%, 2), *Cedrela lilloi* (5.4%, 2), and *Myrcianthes pseudomato* (5.4%, 1), and 5.4% (2) of the nests were found in snags. Compared to the availability of cavities in different tree species, *B. salicifolius*, *J. australis*, and *C. lilloi* were used significantly more than expected ($G_6 = 91.6$, $P < 0.01$). Most nests were found in decay-originated tree cavities (95%, n 5 35), compared to nests excavated (5%, $n = 2$) ($\chi^2 = 21.5$, $P < 0.001$). Cavity location was predominantly in primary branches (43%, $n = 16$), followed by main trunk (32%, $n = 12$), secondary (16%, $n = 6$), and tertiary branches (8%, $n = 3$) ($\chi^2 = 8.9$, $P < 0.03$). Most nests (92%) were found in trees with a DBH >60 cm. Average nest-tree DBH was 89.9 ± 26.9 cm, cavities were located on average at 14.4 ± 3.9 m above the ground, had a horizontal entrance diameter of 13.3 ± 4.5 cm, and a depth of 38.2 ± 38.6 cm. Nests of Tucuman Amazon were found at an elevation range between 1470 and 1710 m a.s.l. Six of the 37 cavities (16%) were reused. One nest-cavity was used in three breeding seasons and five were used twice. One nest-tree had two nest cavities used in separate years. Annual mean nest density in the study area was 0.24 ± 0.04 nests ha-1 ($n = 4$). In two breeding seasons (2006–2007 and 2007–2008) the spatial pattern of active nests was dispersed ($Z = 3.8$ observed distance/expected distance = 1.6 and $Z = 2.7$, observed distance/expected distance = 1.4, respectively), random in 2005–2006 ($Z = -0.3$, observed distance/expected distance = 0.9), and intermediate between random and dispersed in 2008–2009 ($Z = 1.8$, observed distance/expected distance = 1.3). The spatial pattern of all nest-trees used in the four breeding season was clustered ($Z = 4.2$, observed distance/expected distance = 0.6). There was no significant difference in the nearest mean distance among active nests among breeding seasons ($H = 1.4$, $P = 0.69$). For all years combined, the distance to the nearest active nest was significantly greater than the distance between all trees used as nests ($W = 520$, $P < 0.001$). Nesting pairs of Tucuman Amazon were separated by 144.1 ± 152.8 m (range 5 12–674 m), while potential nest-trees were 66.0 ± 55.4 m apart (range = 12–252.4 m). Furthermore, for nests located in 2008–2009, the same pattern prevailed, a nesting pair being significantly farther from the nearest neighbouring pair (138.1 ± 165.3) than from the nearest available nest-tree (53.5 ± 19.5, $Z = 2.29$, $P = 0.02$). The estimation of the density of available suitable cavities for nesting was 4.6 cavities ha-1 (95% CI = 3.1–7.0 cavities ha^{-1}), a coefficient of variation of 20%, and an effective detection width of 9.1 m (95% CI = 7.5–11.1 m).

Nest cavities of Tucuman Amazon were higher, but shallower than nests of other *Amazona* species from the lowlands. This can probably be related to a lower rate of decomposition related to specific tree and sapwood characteristics that compartmentalize decaying wood or are very resistant to fungal decay (Shigo 1984) or to a lower temperature that retards decomposition rates (Politi et al. 2010). As expected, we found that nest cavities of Tucuman Amazon have a larger internal diameter than lowland parrots nests which can be an advantage to maximize nest space and thermal insulation (Joy 2000) in a high elevation breeding habitat where low temperatures are reached. Tucuman Amazon selects three tree species for 83% of nests. Tucuman Amazon sited a high number (95%) of nests in living trees. We reported 0.24 nests ha^{-1} (i.e. one nest every 4 ha) of Tucuman Amazon. Active Tucuman Amazon nests have a dispersed distribution at a mesoscale of 45 ha of the study site. Recently it has been suggested that behavioural spacing requirements of nesting parrots may limit breeding densities and restrict management strategies to increase numbers of nesting pairs within protected areas (Salinas-Melgoza et al. 2009). Tucuman Amazon shows significantly shorter distances among nearest nest-trees than distance among nearest breeding pairs in a breeding season suggesting that breeding pairs influence spacing of conspecifics. The fact that we found a stable number of nests during the four breeding seasons for the fixed area under study, that the mean distance to the nearest breeding pairs was similar in every breeding season, and that the spatial pattern of active nests is mainly dispersed, suggests that spacing due to territorial behaviour could be limiting breeding pair density. This limitation may occur despite the availability of suitable cavities. We estimated that approximately 16 suitable cavities were available for each breeding pair (0.25 breeding pair ha^{-1} and four suitable cavities ha^{-1}). Tucuman Amazon occupied 5%

of the suitable cavities available. The relatively low percentage of reuse of nest cavities for Tucuman Amazon may show a strategy for avoiding predation, since predation is the main cause of nest-loss for the species (Rivera 2011). Nests placed in new cavities have lower predation rates compared to nests in previously used cavities (Brightsmith 2005). The frequent shift of nesting sites to avoid predation is probably due to the high density of suitably cavities available for nesting in this mature forest.

7.4 Reproductive Biology of the Tucuman Amazon

We identified a total of 94 breeding attempts for which we recorded nesting success or causes of nesting failure. Data on reproductive output were gathered from 86 nests in which complete clutches were laid (defined as consecutive visits to active nests during which the number of eggs did not change, but the nest remained active). Nests >15 m above the ground were visually inspected using climbing equipment to reach the nest (Perry 1978). During the incubation period nests above 15 m were inspected once only. We monitored contents of nests when females left the nests to be fed by males (Fig. 7.5). When possible, we recorded the date clutches that were initiated, and determined incubation period and length of the entire reproductive effort for each nest. We estimated the following parameters: clutch-size (i.e. number of eggs laid per nesting attempt), brood-size (i.e. number of chicks per nest), hatching success (i.e. proportion of eggs present in the nest at the end of incubation that hatch), fledging success (i.e. proportion of nestlings that fledge), and nesting success (i.e. proportion of nests producing at least one fledgling). To allow comparison with other studies, reproductive output was expressed as fledglings per successful nest and fledglings per laying female (including successful and unsuccessful females). We considered a nest successful when it produced at least one fledgling. We considered a nest abandoned when it contained eggs but no adults were recorded in more than 1 h of observation on two successive visits. A nest was considered depredated when all of the eggs disappeared before hatching, nestlings disappeared before reaching 47 days of age, or remains of eggs or nestlings were found in an otherwise empty nest. We used 47 days as the cut-off age because fledglings attain maximum primary feather length at ~50 days and we estimate that before 47 days wings are too short to sustain flight (Rivera 2011). Nests were considered lost by starvation when nestlings were found dead with an empty crop. We examined productivity at different stages of the nesting cycle on the basis of all laying pairs.

We used Mayfield's (1975) method to calculate daily nest-survival rate because this method avoids the bias introduced when nests are found at different stages of the nesting cycle. Because intervals between visits to nests varied we used a maximum-likelihood estimate modification of the Mayfield method (Johnson 1979; Krebs 1989). We calculated the variance of the Mayfield estimator according to Johnson (1979) to make comparisons with the program Contrast (see below). We estimated the daily nest-survival rate (DSR) during the incubation (28 days) and

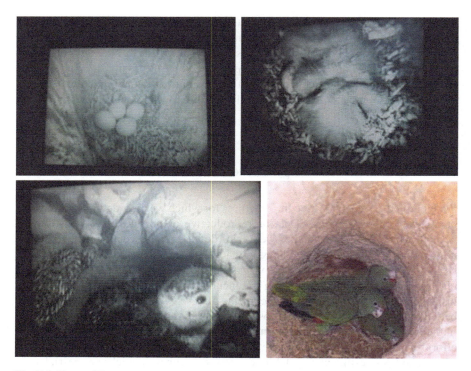

Fig. 7.5 Nests of Tucuman Amazon inspected containing eggs, hatchlings, chicks, and fledglings

nestling (1–50 days after hatching) stages and multiplied the DSR of nests during incubation with the DSR of nests during the nestling stage to obtain the finite survival rate. The program Contrast (Hines and Sauer 1989) was used to compare DSR among periods of the nesting cycle and among breeding seasons. We were not able to use dataset from the 2004–2005 breeding season because DSR had no associated variance. We assumed that a juvenile had fledged when it was absent from a nest at 47 days after hatching. Disappearances at earlier ages were considered to be deaths. In order to understand whether clutch-size of Tucuman Amazon is higher or lower than expected for the body-size of the species, we assessed whether adult Tucuman Amazon body-mass (~280 g, Low 2005) and clutch-size followed the allometric equation ($y = 2.2 + 5.5 \exp.[-0.006x]$) developed by Masello and Quillfeldt (2002) for other species of parrots, where y is clutch-size is and x is body-mass. Additionally, we performed a regression analysis of clutch-size as a function of body-mass for all species of *Amazona* parrots for which we could obtain data, to test if clutch-size of Tucuman Amazon lies inside the 95% confidence interval of the allometric relationship.

In general, in parrots of the genus *Amazona* the female brood the eggs and the male provide food to the incubating female (Low 2005). In Tucuman Amazon the male provide food to the incubating female twice a day, early in the morning and

7 Nesting Ecology of the Tucuman Amazon (*Amazona tucumana*) in the Cloud Forest... 143

late in the afternoon. Females pass less than 10 min outside the nest when they receive food from males (Rivera et al., unpublished). Females leave the nest 15 days after hatching of the chicks. After the female stop brooding both parents feed the nestlings three times per day adding a visit to the nest at noon (Rivera et al., unpublished). Recently hatched chicks weight 12 g, coincident with records in captivity (Low 2005). Length of wing, tail, tarsus, and bill were lower for fledglings than for adults if we compare our data with those of Forshaw and Cooper (1989) for 19 adults Tucuman Amazon. Our measurements were done 1–2 days before fledging suggesting that fledglings are in development and growing when leaving the nests. Weight of fledglings (about 265 g) is similar to chicks of 70 days raised in captivity (260–269 g) and lower to the weight of adults (280 g) (Low 2005). Because sexes are alike in Tucuman Amazon the only form to determine sex is by ADN analysis. Our analysis using two conserved genes (chromo-helicase-DNA-binding) located in the sexual chromosome (Griffiths et al. 1998) showed no significant differences ($\chi2 = 1.8$, gl = 1, $P = 0.18$) (42.5% males and 57.5% females for 40 chicks of 12 nests). The length of the incubation period (mean 28.33 ± 0.58 days, range = 28–29), nestling period (49.7 ± 1.1 days, range = 49–51), and overall nesting period of Tucuman Amazon were determined from data from three nests for which date of laying of the first egg, date of hatching of the first egg, and date of fledging of the first chick were known with certainty. Clutch-size ranged from one to five eggs (mean 3.6 ± 1.0, mode = 4, $n = 86$). Clutches of one are most likely to be complete clutches because we did not find any evidence of partial loss or predation. The smallest overall clutch-size was recorded during the 2008–2009 breeding season, and differed significantly from mean annual clutch-size of three other breeding seasons ($H4 = 9.6$, d.f. = 4, $P = 0.03$). Overall hatching success (proportion of eggs that hatch) was 0.77, with the lowest rate (0.54), recorded in 2006–2007. The number of fledglings per successful nest (overall mean 3.2 ± 0.2, $n = 51$) differed among years with lowest values in 2005–06 and 2006–07 ($H4 = 8.3$, d.f. = 4, $P = 0.05$). The number of fledglings per laying female was 2.3 ± 0.8 ($n = 86$) and differed significantly among years, with the lowest value during the 2006–2007 breeding season ($H4 = 11.9$, d.f. = 4, $P = 0.01$). The daily survival rate during the incubation period did not differ significantly among years. The daily survival rate during the nestling period and the entire nesting cycle differed significantly among breeding seasons. The highest finite survival rate for the nestling period and the nesting cycle was in the 2005–2006 breeding season and the lowest finite survival rate was in 2006–2007. Over the nesting cycle, the primary causes of nest-loss were predation (16%) and abandonment (12%). Nine nests were depredated during the incubation period and six during the nestling period. Abandonment was the main cause of failure during the incubation period, followed by predation, whereas predation was the main cause of failure during the nestling period. Nesting failure was higher during incubation: 22 nests were lost during this period (63% of all losses), whereas 13 nests were lost during the nestling period (37%). Eleven nests were abandoned during incubation, whereas none was abandoned during the nestling period. In two breeding seasons (2006–2007 and 2008–09) predation was the primary cause of nest-loss, in one breeding season (2005–2006) the main cause was abandonment, and in one season

(2007–2008) predation and abandonment were equal causes. Predation of four nests <100 m apart might be attributed to Black-capped Capuchin monkeys (*Cebus apella*) that were observed on a tree in which we had confirmed active nests the previous day and found empty the day after the monkeys were observed. In two nests with advanced nestlings we found bundles of plucked feathers at the entrance to the nesting cavity and on the ground, suggesting predation by mammals. Most nesting failures ($n = 13$) occurred during the 2006–2007 breeding season, with fewer failed attempts per year in the other breeding seasons.

With a mean body-mass of 280 g, the predicted clutch-size of Tucuman Amazon would be 3.2 eggs based on the allometric equation of Masello and Quillfeldt (2002). The 95% confidence interval for the expected clutch-size of the allometric equation for 14 other species of *Amazona* parrots is 2.94–3.43, which is significantly lower than the mean clutch-size of 3.6 that we observed for the Tucuman Amazon. The Tucuman Amazon has high rates of nesting success, large clutches, and a large number of fledglings per laying female. These results differ from trends observed in other bird species that tend to shift to a slower life-history strategy with increased elevation (Sandercock et al. 2005a, b). There are several alternative explanations for our finding that the Tucuman Amazon has the second largest clutch-size among *Amazona* parrots. Firstly, the health (or nutritional state) of female birds, which is strongly related to availability of food, might affect both the number and quality of the eggs laid (Lack 1954; Martin 1987); secondly, birds with higher rates of nesting success may be expected to lay larger clutches (Skutch 1985); thirdly, clutch-size is inversely related to population density (Ricklefs 1980), which, in the case of nesting Tucuman Amazon, may be low; and lastly, the large clutch-size may be evidence of a life-history trade-off in which increased productivity is compensated for by lower survival of juveniles or adults (Bears et al. 2009). The overall mean number of fledglings per breeding female and the overall nesting success for Tucuman Amazon calculated as the maximum-likelihood estimator are the second highest values reported for *Amazona* parrots. Likewise, the Tucuman Amazon has the lowest loss of initial reproductive investment among *Amazona* parrots that have been studied. The probability of nesting success varied significantly among years at the combined nestling stages and showed significant differences among years in nesting success over the entire nesting cycle. Inter-annual variability in the availability of key food items may influence productivity and nesting survival of Tucuman Amazon, as is the case for the Lilac-crowned Parrot (Renton and Salinas Melgoza 2004), the Austral Parakeet (*Enicognathus ferrugineus*) (Díaz et al. 2012), and for many other bird species (Newton 1980).

We recorded 11 nests with intact eggs that were abandoned, although we did not check for infertility or diseases. Low temperatures (e.g. 8C minimum on rainy days) in the cloud forest might make Tucuman Amazon more susceptible to loss of eggs through embryonic chilling (Stoodley and Stoodley 1990), especially when incubating adults are disturbed. Factors that could explain abandonment are predation of incubating adults or food scarcity, which might lead females to end incubation because they cannot fulfil the high energy requirements of incubation. Predation and abandonment of nests accounted for 28% of all losses of nests in the Tucuman

Amazon. Because Tucuman Amazon inhabit the cloud forest between 1400 and 2200 m above sea level, where richness and abundance of predators are expected to be lower than at lower elevations (Skutch 1985), we expected to observe a lower rate of nest predation than for other *Amazona* species of lowlands. Although there are no snakes—one of the most important predators of other *Amazona* parrots (Enkerlin-Hoeflich 1995; Renton and Salinas Melgoza 2004; Berkunsky et al. 2011)—in cloud forests of the Southern Yungas, other predators present include Black-capped Capuchins, at least four species of raptor (Barred Forest-Falcon, *Micrastur ruficollis*; White-rumped Hawk, *Parabuteo leucorrhous*; Roadside Hawk, *Buteo magnirostris*; Bicoloured Hawk, *Accipiter bicolor*) and three other potential nest predators: two mammals, the Tayra (*Eira barbara*) and the Lesser Grison (*Galictis cuja*), and the Plush-crested Jay (*Cyanocorax chrysops*). We frequently observed all of these species in the breeding habitat of Tucuman Amazon. The Tucuman Amazon breeds during the rainy season and, therefore, we expected to find a high number of flooded nests. However, we found few flooded nests in extremely rainy years (i.e. 2006–2007 and 2007–2008).

7.5 Conservation Implications

Cloud forest in the Southern Yungas is subject to intense selective logging, primarily of one species (i.e., *Cedrela lilloi*). Until recently (late 1990s) *Podocarpus parlatorei* was intensively logged for paper pulp production, but now paper pulp is made from sugar cane. Cloud forests are also frequently replaced by grasslands through intentional fires set to facilitate extensive cattle grazing (Brown and Grau 1993; Brown et al. 2001). The probability of encountering a cavity increases with tree age (Newton 1994); therefore, the surplus of suitable cavities found in this study for each breeding Tucuman Amazon pair in an old-growth forest is not surprising. In managed forests, the availability of suitable cavities for nesting might be lower (Newton 1994; Politi et al. 2010), and if Tucuman Amazon requires a certain number of suitable cavities in their home range, it is expected that the species will be particularly vulnerable to the loss of nesting habitat through the impacts of selective logging or habitat modification (Monterrubio-Rico et al. 2009). Considering that most of the Southern Yungas is under timber exploitation it is reasonable to expect that this might be the reason that Tucuman Amazon has not recovered (Rivera et al. 2010); i.e. large trees with cavities are probably lacking. To ensure the conservation of Tucuman Amazon outside protected areas it is necessary that forest management guidelines promote the retention of large *B. salicifolius* trees, since this species was used more frequently than would be expected from its abundance. This might be possible because this species does not have a high timber value. However, this is probably more difficult with other tree species (such as *C. lilloi* and *J. australis*) that are selected for nesting by Tucuman Amazon but have high timber values. Finally, the nesting and spatial requirements of Tucuman Amazon could limit management actions intended to increase the density of nesting pairs. A mean distance among suitable cavities of at least 150 m could represent the minimum distance to consider in the spacing of active pairs to avoid exclusion by other pairs especially if nest box provision is necessary.

7.6 Knowledge Gaps

We identified the following aspects that deserve future attention from a research perspective to have a complete understanding of the ecology or Tucuman Amazon:

1. Diet and food availability.
2. Survival of fledglings and adults.
3. Genetic structure of populations in North-south latitudinal gradient.
4. Movement ecology studying daily and seasonal movements and dispersion of juveniles.

References

Aitken KEH, Martin K (2004) Nest cavity availability and selection in aspen conifer groves in a grassland landscape. Can J For Res 34:2099–2109

Berkunsky I, Reboreda JC (2009) Nest site fidelity and cavity reoccupation by blue-fronted parrots *Amazona aestiva* in the dry Chaco of Argentina. Ibis 151:1–34

BirdLife International (2018) *Amazona tucumana*. The IUCN Red List of Threatened Species 2018: e.T22686246A93104452. https://doi.org/10.2305/IUCN.UK.2016-3.RLTS. T22686246A93104452.en. Downloaded on 01 August 2019

Bond J, Meyer de Schauensee R (1943) The birds of Bolivia. Part II. Proceeding of the Academy of Natural Sciences of Philadelphia 95:167–221

Arias M, Bianchi AR (1996) Estadísticas climatológicas de la provincia de Salta. EEA Salta. Dirección de Medio Ambiente y Recursos Naturales, Gobierno de Salta, Salta

Auer SK, Bassar RD, Fontaine JJ, Martin TE (2007) Breeding biology of passerines in a subtropical montane forest in northwestern Argentina. Condor 109:321–333

Bears H, Martin K, White GC (2009) Breeding in high-elevation habitat results in shift to slower life-history strategy within a single species. J Anim Ecol 78:365–375

Berkunsky I, Kacoliris F, Faegre S, Ruggera R, Carrera J, Aramburú R (2011) Nest predation by tree-snakes on cavity nesting birds in dry Chaco woodlands. Ornitologia Neotropical 22:459–464

Brightsmith DJ (2005) Competition, predation and nest niche shifts among tropical cavity nesters: ecological evidence. J Avian Biol 36:74–83

Brown AD, Grau HR, Malizia L, Grau A (2001) Argentina. In: Kappelle M, Brown AD (eds) Bosques nublados del Neotrópico. InBio, Santo Domingo de Heredia

Brown AD, Grau HR (1993) Investigación, conservación y desarrollo en las selvas subtropicales de montaña. LIEY, Universidad Nacional de Tucumán, Argentina, Tucumán

Brown AD (1995) Fenología y caída de hojarasca en las selvas montanas del Parque Nacional El Rey, Argentina. In: Brown AD, Grau HR (eds) Investigación, conservación y desarrollo en las selvas subtropicales de montaña. LIEY, Universidad Nacional de Tucumán, Argentina, Tucumán

7 Nesting Ecology of the Tucuman Amazon (*Amazona tucumana*) in the Cloud Forest... 147

Buckland ST, Anderson DR, Burnham KP, Laake JL, Borchers DL, Thomas L (2001) Introduction to distance sampling; estimating abundance of biological populations. Chapman and Hall, New York

Burnham KP, Anderson DR (2002) Model selection and multimodel inference: a practical information-theoretic approach, 2nd edn. Springer Verlag, New York

Cabrera AL, Willink A (1980) Biogeografía de América Latina. Organization of American States, Washington

Cabrera AL (1994) Regiones Fitogeográficas Argentinas. Enciclopedia Argentina de Agricultura y Jardinería. Tomo II. Editorial ACME S.A.C.I, Buenos Aires

Díaz S, Kitzberger T, Peris S (2012) Food resources and reproductive output of the austral parakeet (*Enicognathus ferrugineus*) in forests of northern Patagonia. Emu 112:234–243

Enkerlin-Hoeflich EC (1995) Comparative ecology and reproductive biology of three species of Amazon parrots in northeastern Mexico. Doctoral thesis. College Station, Texas A&M University

Fjeldså J, Krabbe N (1990) Birds of the high Andes. Apollo Books, Stenstrup

Forshaw JM (1977) Parrots of the world, 3rd edn. Lansdowne Press, Willoughby

Forshaw JM, Cooper WT (1989) Parrots of the world. Lansdowne Editions, Sydney

González Elizondo JJ (1998) Productividad, causas de mortalidad en nidos y dieta de los polluelos de tres especies de loro del género *Amazona* en el sur de Tamaulipas. Tesis MSc. Universidad del Noroeste de México

Griffiths R, Double MC, Orr K, Dawson R (1998) A DNA test to sex most birds. Mol Ecol 7:1071–1075

Hines JE, Sauer JR (1989) Program CONTRAST: a general program for the analysis of several survival or recovery rate estimates. Fish and Wildlife Technical Report 24:1–7

Hoy G (1968) Uber brutbiologie und eier einiger vogel aus nordwest-Argentina. J Ornithol 109:425–433

Hueck K (1978) Los Bosques de Sudamérica. Ecología, composición e importancia económica. Agencia Alemana de Cooperación Técnica (GTZ), Berlín

Hunter ML Jr, White AS (1997) Ecological thresholds and the definition of old-growth forest stands. Nat Areas J 17:292–296

Ibisch PL, Mérida G (2003) Biodiversidad: La riqueza de Bolivia. Estado de conocimiento y conservación. Ministerio de Desarrollo SostenibleEditorial FAN, Santa Cruz de la Sierra

Johnson DH (1979) Estimating nest success: the Mayfield method and an alternative. Auk 96:651–661

Joy JB (2000) Characteristics of nest cavities and nest trees of the red-breasted sapsucker in coastal montane forests. J Field Ornithol 71:525–530

Juniper P, Parr M (1998) Parrots: a guide to parrots of the world. Yale University Press, New Haven

Koenig SE (2001) The breeding biology of black-billed parrot *Amazona agilis* and yellow-billed parrot *Amazona collaria* in cockpit country, Jamaica. Bird Conservation International 11:205–225

Kappelle M, Brown AD (2001) Bosques nublados del neotrópico. INBIO, Costa Rica

Krebs CJ (1989) Ecological methodology. Harper Collins, New York

Krebs CJ (1999) Ecological methodology, 2nd edn. Benjamin Cummings, Menlo Park

Lack D (1954) The natural regulation of animal numbers. Clarendon Press, Oxford

Martin TE (1987) Food as a limit on breeding birds: a life-history perspective. Annu Rev Ecol Syst 18:453–487

Low R (2005) Amazon parrots, aviculture, trade, and conservation. Dona/Insignis Publications, Prague

Manly BFJ, Mcdonald LL, Thomas DL, Mcdonald TL, Erickson WP (2002) Resource selection by animals: statistical design and analysis for field studies. Kluwer Academic Publishers, Boston

Martinez J, Prestes NP (2008) Biologia da conservacao: estudo de caso com papagaio-charao e outros papagaios brasileiros. Editora UPF, Passo Fundo

Masello J, Quillfeldt P (2002) Chick growth and breeding success of the burrowing parrot. Condor 104:574–586

Mayfield H (1975) Suggestions for calculating nest success. Wilson Bulletin 84:456–466

Mendoza EA (2005) El clima y la vegetación natural. In: Minetti JL (ed) El clima del noroeste Argentino. Ed. Magna, Tucumán

Mitchell A (2005) The ESRI Guide to GIS Analysis. Volume 2. ESRI, Redlands

Monterrubio-Rico TC, Ortega-Rodríguez JM, Maríntogo MC, Salinas-Melgoza A, Renton K (2009) Nesting habitat of the lilac-crowned parrot in a modified landscape in Mexico. Biotropica 41:361–368

Newton I (1994) The role of nest sites in limiting the numbers of hole nesting birds: a review. Biol Conserv 70:265–276

Narosky T, Yzurieta D (2003) Aves de Argentina y Uruguay: guía para la identificación. Edición de oro, 15a ed. Vasquez-Mazzini, Buenos Aires, Argentina

Newton I (1980) The role of food in limiting bird numbers. Ardea 68:11–30

Orfila RN (1938) Los psittaciformes argentinos (cont). Hornero 7:1–21

Perry DR (1978) A method of access into the crowns of emergent and canopy trees. Biotropica 10:155–156

Pidgeon AM, Rivera L, Martinuzzi S, Politi N, Bateman B (2015) Will representation targets based on area protect critical resources for the conservation of the Tucuman parrot? The Condor Ornithological Applications 117:503–517

Politi N, Rivera L (2005) Abundance and distribution of parrots along the elevational gradient of Calilegua National Park, Argentina. Ornitología Neotropical 16:43–52

Politi N, Hunter M Jr, Rivera L (2010) Availability of cavities for avian cavity nesters in selectively logged subtropical montane forests of the Andes. For Ecol Manag 260:893–906

Renton K (2000) Scarlet macaw. In: Reading RP, Miller B (eds) Endangered animals: a reference guide to conflicting issues. Greenwood Press, Westport

Renton K, Salinas Melgoza A (2004) Climatic variability, nest predation, and reproductive output of lilac-crowned parrots (*Amazona finschi*) in tropical dry forest of western Mexico. Auk 121:1214–1225

Richardson DM, Bradford JW, Range PG, Christensen J (1999) A video probe system to inspect red-cockaded woodpecker cavities. Wildl Soc Bull 27:353–356

Ricklefs RE (1980) Geographical variation in clutch size among passerines birds: Ashmole's hypothesis. Auk 97:38–49

Rivera L (2011) Ecología, biología reproductiva y conservación del Loro alisero (*Amazona tucumana*) en Argentina. Tesis Doctoral. Córdoba, Argentina: Universidad Nacional de Córdoba

Rivera L, Politi N, Bucher EH (2007) Decline of the Tucuman parrot *Amazona tucumana* in Argentina: present status and conservation needs. Oryx 41:101–105

Rivera L, Politi N, Bucher EH (2012) Nesting habitat of the Tucuma'n parrot *Amazona tucumana* in an old-growth cloud forest of Argentina. Bird Conservation International 22:398–410

Rivera L, Politi N, Bucher EH, Pidgeon AM (2013) Nesting success and productivity of Tucuman parrots (*Amazona tucumana*) in high-altitude forests of Argentina: do they differ from lowland *Amazona* parrots? Emu 114:41–49

Rivera L, Rojas Llanos R, Politi N, Hennessey B, Bucher EH (2010) The near threatened Tucumán parrot *Amazona tucumana* in Bolivia: insights for a global assessment. Oryx 44:110–113

Rocha AV, Rivera L, Martinez J, Prestes NP, Caparroz R (2014) Biogeography of speciation of two sister species of Neotropical *Amazona* (Aves, Psittaciformes) based on mitochondrial sequence data. PLoS One 9:1–10

Russello MA, Amato G (2004) A molecular phylogeny of *Amazona*: implications for Neotropical parrot biogeography, taxonomy, and conservation. Mol Phylogenet Evol 30:421–437

Salinas-Melgoza A, Salinas-Melgoza V, Renton K (2009) Factors influencing nest spacing of a secondary cavity-nesting parrot: habitat heterogeneity and proximity of conspecifics. Condor 111:305–313

Sandercock BK, Martin K, Hannon SJ (2005a) Life history strategies in extreme environments: comparative demography of arctic and alpine ptarmigan. Ecology 86:2176–2186

Sandercock BK, Martin K, Hannon SJ (2005b) Demographic consequences of age-structure in extreme environments: population models for arctic and alpine ptarmigan. Oecologia 146:13–24

Saunders DA, Smith G, Rowley I (1982) The availability and dimensions of tree hollows that provide nest sites for cockatoos in Western Australia. Australian Wildlife Research 9:541–556

Shigo AL (1984) Compartmentalization: a conceptual framework for understanding how trees grow and defend themselves. Annu Rev Phytopathol 22:189–214

Snyder NFR, Wiley JW, Kepler CB (1987) The parrots of Luquillo: natural history and conservation of the Puerto Rican parrot. Western Foundation of Vertebrate Zoology, Los Angeles

Skutch AE (1985) Clutch size, nesting success, and predation on nests of Neotropical birds, reviewed. Ornithol Monogr 36:575–594

Stoodley J, Stoodley P (1990) Genus *Amazona*. Avian Publications, Altoona

Tella JL, Rojas A, Carrete M, Hiraldo F (2013) Simple assessments of age and spatial population structure can aid conservation of poorly known species. Biol Conserv 167:425–434

Thomas L, Laake JL, Strindberg S, Marques FFC, Buckland ST, Borchers DL, Anderson DR, Burnham KP, Hedley SL, Pollard JH, Bishop JRB, Marques TA (2006) Distance 5.0 release 2. Research unit for wildlife population assessment. University of St. Andrews, St Andrews

Tortorelli L (1956) Maderas y bosques Argentinos. Acme Agency, Buenos Aires

Wetmore A (1926) Observations on the birds of Argentina, Paraguay, Uruguay and Chile. Bulletin of the United States National Museum 133:1–448

Chapter 8
Adaptive Strategies of Frugivore Bats to Andean Cloud Forests

Adriana Ruiz and Pascual J. Soriano

8.1 Some Historic Considerations

It is well known that first feature distinguishing bat assemblages from along an altitudinal gradient is the drastic reduction in the number of constituent species (Graham 1983; Patterson et al. 1996). In two previous papers, we examined how the bat assemblages of tropical cloud forests of Venezuelan Andes were composed and structured (Soriano 2000; Soriano et al. 1999) and we demonstrate that use of functional groups, such as trophic categories, is more useful for ecological interpretation of differences in bat assemblage structure than the simple comparison of lists of names; thereby, we introduce the concept of "Trophic Equivalent" (TE) as a way to make functional comparisons among bat assemblages (Soriano 2000).

Andean cloud forest bat assemblages from Venezuela show two relevant characteristics, distinguishing them from those of lowland rainforests: (1) Trophic simplification: The importance values for each functional category showed that the assemblage is structured mainly on the base of frugivore and insectivore diets with a small contribution from nectarivores. The rest of the categories are either not represented or their contribution is very small, such as hematophages, whose presence in the cloud forests is associated with disturbed areas used for cattle farming. (2) Trophic segregation of body sizes: The species with lowest body mass were largely insectivores, while the inverse relationship was true for the frugivores. The nomadic frugivore species were all large (in size), and the species in the sedentary frugivore

---

A. Ruiz
Postgrado en Ecología Tropical (ICAE), Facultad de Ciencias, Universidad de Los Andes, Mérida, Venezuela

P. J. Soriano (✉)
Laboratorio de Ecología Animal, Departamento de Biología, Facultad de Ciencias, Universidad de Los Andes, Mérida, Venezuela
e-mail: pascual@ula.ve

© Springer Nature Switzerland AG 2021
R. W. Myster (ed.), *The Andean Cloud Forest*,
https://doi.org/10.1007/978-3-030-57344-7_8

guild, which were small in size, in reality came from nectarivore guild but which occasionally ate fruits.

In the bat assemblages of Andean cloud forests, frugivores represent the most species-rich guild, in contrast with lowland rain forest bat communities, where insectivores are the dominant guild (Graham 1983; Patterson et al. 1996; Soriano 2000; Soriano et al. 1999). Apparently, insectivores of tropical origin have a limit to their vertical distribution that prevents them from accessing Andean cloud forests and only some Vespertilionidae of Neartic origin reach these forests, along with a few representatives of the Molossidae (Soriano 2000; Soriano et al. 1999). Variations in metabolism observed in bats are mainly related to body mass and feeding habits: bats with insectivorous or hematophagous diets have low rates of metabolism; bats with combined diets (frugivore–carnivore) have low-to-intermediate rates of metabolism; frugivores have high rates of metabolism; and metabolic rates for nectarivores are very high. Associated with basal metabolic rate is the capacity to regulate body temperature; therefore, within the same feeding habit, large bats regulate their temperature better than small ones. Similarly, for bats of a given body mass, insectivores maintain lower body temperatures and are more dependent on environmental temperature variations than frugivores (Bonaccorso and McNab 1997; McNab 196 9,1970,1982a,1983,1986,1989). At altitudes between 2000 and 3000 m in the tropical Andes, the mean temperature is 7–10.8 °C lower than in the lowlands (Sarmiento 1986). By virtue of their lower temperatures, high mountain environments demand higher energetic expenditures for endotherms to maintain a constant body temperature. Therefore, for endotherms, the low temperatures of montane environments may impose important physiological constraints on the possession of an adequate energetic balance.

Such constraints seem to occur in members of the Emballonuridae, Mormoopidae, Thyropteridae, Furipteridae, and Natalidae families, which never or rarely are found at elevations above 1000 m (Graham 1983; Patterson et al. 1996; Soriano 2000; Soriano et al. 1999). Successful adaptation to cold environments presupposes modifications of physiological responses characterizing high mountain species. Among such adaptive modifications are one or several of the following physiological features: increase of basal metabolic rate, displacement of thermoneutrality zone to a lower temperature range, decrease of thermal conductance by compensatory increase of insulation, and daily facultative or obligatory torpor. We tested the hypothesis that bats of the Andean highlands showed distinctive metabolic responses compared with bats from lowland forests (Soriano et al. 2002); additionally, we compared the existing literature with new information on three bat species having the following food habits: a nectarivore (*Anoura latidens*), a frugivore (*Sturnira erythromos*), and an insectivore (*Tadarida brasiliensis*). Basal metabolic rate, as determined by oxygen consumption, thermal conductance, and body temperature were measured at ambient temperatures of 10–38.8 °C. Some distinctive metabolic responses of these bat species, although varying with respect to food guild, allowed us to separate them from counterpart species that are typically found in lowland forests. *A. latidens* was characterized by higher basal metabolic rate; however, thermal conductance and lower critical temperature values did not showed an adaptation

8 Adaptive Strategies of Frugivore Bats to Andean Cloud Forests

to cool environments, and as expected. *S. erythromos* also increased its basal metabolic rate, but it maintained thermal conductance as expected, which implied a very important displacement of thermoneutral zone to lower temperatures. At temperatures below lower critical temperature, in addition to an endothermic response, *S. erythromos* sometimes expressed a hypothermic response or facultative torpor, independent of sex and body mass. *T. brasiliensis* was a lower basal metabolic rate and thermal conductance and also was its thermoneutral zone range displaced toward lower temperatures. Likewise, this species entered obligate torpor when ambient temperatures were below 22.8 °C.

8.2 Frugivore Bats of Genus *Sturnira*: A Study Case

The fruit bats (family: Phyllostomidae) inhabit mainly in the Neotropics, and few species, belonging mainly to the subfamily Stenodermatinae, occur marginally in the subtropical zones of both in North and South America (McNab 1982a,b; Simmons 2005). In tropical highlands, fruit bats of the genera *Platyrrhinus* and *Sturnira* usually inhabit elevations above 2000 m (Alberico et al. 2000; Graham 1983; Koopman 1978; Patterson et al. 1996; Soriano et al. 1999; Tuttle 1970). However, only the second genus is diverse at high altitudes (de la Torre 1961; Iudica 2000). The distribution of frugivorous bats in the tropical region has been considered a result of the combination of historical (phylogeny) and ecological characteristics that would explain its current restricted range (Colwell and Lees 2000; Lomolino 2001; McNab 1982a; Rahbek 1995; Stevens 2006). The altitudinal limits of some species that inhabit the Andean region do not appear to be correlated with changes in vegetation and productivity along altitudinal gradients (Graham 1983,1990; Terborgh 1971,1977; Willig et al. 2003). In comparison with other Neotropical bat families, the fruit-eating phyllostomids show thermoregulatory and food habit restrictions that affect their distribution limits (McNab 1969,1980a, b,1988; Stevens 2004,2006).

The physiologic mechanisms of ecological and evolutionary relevance are tools that allow us to know how some physiological trials can have an effect on the distribution patterns (McNab 1969,1973,1974,1976,1980a, b,1982a,2003,Soriano et al. 2002; Stevens 2004; Willig and Bloch 2006). A high-energy intake has been associated with species that are good thermoregulators as a consequence of their higher metabolic rates (McNab 2003), with the exception of small size species, whose high metabolic rates can be insufficient to compensate their higher heat loss (Audet and Thomas 1997; McNab 1982a; Soriano et al. 2002). The apparent inability of fruit bats to enter in torpor (hibernation) would presumably restrict them to occur preferably in the tropical region (McManus 1977), such as those in which flying foxes (Pteropodidae) have been registered. Some pteropodid bats that inhabit in lowland and subtropical zones use daily torpor due to the low resources availability (Bartels et al. 1998; Bonaccorso and McNab 1997; Coburn and Geiser 1998; Geiser et al. 1996; Law 1994; McNab 1989). In Andean highlands, ambient temperatures decrease by 0.6 °C with every 100 m increase of elevation (Sarmiento 1986); therefore, fresh fruit availability and productivity diminish, both factors should affect bat species diversity along of the altitudinal gradient (Graham 1983; Rosenzweig 1992; Terborgh 1971; Tilman 1982).

The members of the genus *Sturnira* are strict frugivores and show a widespread distribution in the Western Hemisphere (Simmons 2005). Almost half of the 14 species of this genus reach highlands in tropical regions, while others are restricted to lowlands or they may reach the mountains marginally (Contreras-Vega and Cadena 2000; de la Torre 1961; Simmons 2005). The small differences in body size along with morphological similarities could explain the use of similar resources in highlands (Giannini 1999; Molinari and Soriano 1987; Ruiz 2006). So far, very little is known concerning the thermoregulation capacity of high mountains species; the only species in which mechanism has been reported until know is *Sturnira erythromos*, in which higher basal metabolic rate associated with small body size (15 g) appears to cause difficulties in metabolic maintenance of body temperatures. Therefore, these limitations are compensated with daily torpor (Soriano et al. 2002). Although torpor is not an exclusive feature of high altitude dwellers, it could be used as an energy-saving trait during resource shortage periods (Geiser and Coburn 1999; Geiser et al. 1996; McNab and Bonaccorso 2001).

In frugivorous highland bats, small body size combined with high mass-specific energy expenditure could result in a limited ability to store vital resources (McNab 1982a). If physiological limits depend on body mass (McNab 1983), it is feasible that inter-and intra-specific physiological differences between small and large species of *Sturnira* (15–50 g) could be explained by means of body mass. Other fruit bats that reach marginally the Andean mountains such as *Artibeus jamaicensis*, *Carollia perspicillata*, and *Sturnira lilium* show similar heterothermic responses; yet in these cases, resource shortage must be the decisive factor that determines their higher distribution limits (Audet and Thomas 1997; Studier and Wilson 1970). It is difficult to separate the effect of resource shortage from temperature variation along the altitudinal gradient. The fruits consumed by these species are distributed along the Andean slopes, and their abundance increases in cloud forests (Giannini 1999; Ruiz 2006; Soriano 1983). In some cloud forests, as many as five different species of *Sturnira* can coexist (Soriano 1983). If the fruit availability is not restrictive at high elevations for these species, low temperature is the most probable factor that conditions these species altitudinal distribution, as previous studies suggest (Soriano 1983). Is it possible that these species are capable of entering torpor as their congener *S. erythromos* and as other frugivorous and nectarivorous bats of old world?

The aim of this research was to answer two questions: (1) what is the effect of low temperatures on the thermoregulation of highlands fruit bats? and (2) what are the physiological mechanisms that allow them to colonize and to establish in the Andean highlands? We selected three species of the genus *Sturnira* that inhabit cloud forests of the Venezuelan Andes: *Sturnira bidens*, *S. bogotensis*, and *S. ludovici* (>2000 m of elevation; Linares 1998). In this ecosystem, these species are sympatric with *S. erythromos* (Ruiz 2006; Soriano 1983). The physiological traits

8 Adaptive Strategies of Frugivore Bats to Andean Cloud Forests

(regulation of the body temperature, basal metabolic rate, and humid thermal conductance) were compared with the available information concerning lowlands frugivorous bats (Family Phyllostomidae). Finally, we analyzed the ecological factors associated with the physiological measurements and we discussed the physiological consequences of thermoregulation on the altitudinal distribution patterns of the genus *Sturnira*.

8.3 Materials and Methods

Study animals: The three species of the genus *Sturnira* (*S. bidens*, *S. bogotensis*, and *S. ludovici*) were captured in an Andean cloud forest near of the city of Merida, Merida State, Venezuela, between 2000 and 2500 m altitude. We use mist nets to capture bats at foraging sites in two Andean localities: (1) Monterrey (71° 7'W; 8° 41'N; to 8 km NE Merida; at 2300 m) and (2) Monte Zerpa (71° 10'W; 8° 39'N; to 4 km NNE Merida; at 2000–2200 m). Once captured, we selected only the adults bats and confirmed that the females did not show signs of reproductive activity (pregnancy or lactation); then animals were placed in individual cloth bags and transported within 0.5–1 h to the Laboratory of Animal Ecology (Universidad de Los Andes, Merida, Venezuela). We maintained them in captivity for 3 or 4 days, under controlled conditions of light:darkness (12 L:12D) and ambient temperature (among 21–23 °C). During this period, we feed the bats with commercial baby food of tropical fruits 4–6 h prior measurements, with the purpose of assuring that animals were post-absorptive (Kovtun and Zhukova 1994). We followed guidelines for procedures related to animal care and use, approved by the American Society of Mammalogists (Gannon et al. 2007).

Experimental protocol: Our field and laboratory work were conducted between November 2002 and July 2005. We performed the experiments between 0900 and 1800 h, when the bats were at rest and post-absorptive. We measured experimentally the oxygen consumption (VO_2 in mL O_2 g^{-1} h^{-1}) and humid thermal conductance (C in mL O_2 g^{-1} h^{-1} $°C^{-1}$) in an open-circuit respirometry system, modifying ambient temperature from 10 to 40 °C. We placed each bat in sealed plastic-PVC chamber (450 and 900 mL) with a wire mesh inside from which the bats could hang in a natural posture. We submerged the sealed chamber in a thermoregulated water bath. After an acclimation period of ca. 60 min, the current oxygen levels were measured for 2–4 h. For individuals at ambient temperature of 34 °C or more, the test period was reduced to 30 min. Measurements of air temperature inside the chamber were monitored with a copper–constantan thermocouple that passed through the stopper and connected to a Microprocessor Digital Thermometer (Model HH23 Omega, Stanford, Connecticut). The lid chamber had two ports connected to tubes for air entry and exit. Room air was drawn in by a pump at a flow rate that we regulated between 100 and 140 or 220 and 260 mL/min for the 450 and 900 mL chambers, respectively. These airflow rates inside the chambers were adequate, since metabolic rates were not affected by oxygen flow rates. Exiting from the chamber, carbon dioxide and water were removed from the air by color-indicating soda lime and silica gel, respectively, and then flow rate was measured by a flow meter Cole Parmer (Model P/N: 10130).

We measured the oxygen concentration of the air with an oxygen analyzer (Applied Electrochemistry, model S3A-II Ametek, Pittsburg, Pennsylvania), which we calibrated previously. Voltage output of the analyzer was recorded with a personal computer, using A/D converter card, and acquisition software (created by the Laboratory of Scientific Instrumentation, Universidad de Los Andes, Merida). Bats were removed from the experimental chamber after assuring that oxygen concentration was minimal and stable for at least 30 min. We measured body temperature (Tb) of the animals immediately before and within 15 s after using rectal insertion of a fine (0.2 mm diameter) thermocouple probe calibrated to the nearest 0.1 °C, which was read with a digital thermometer. Bats were considered torpid when body temperature fell below 30 °C and when those rewarmed reached their initial body temperature. Animals were weighed before and after the experiments, and a linear decrease of body mass throughout the experiment was assumed for calculation of mass-specific metabolic rate. We took care that the individuals did not lose more than 10% of their body mass.

Estimate of parameters: We tested the effect of ambient temperature (Ta) on physiological traits using lineal regressions. We corrected the values to standard pressure and temperature, and we calculated the rate of metabolism from Dépocas and Hart (1957) and the thermal conductance "humid" (C) with the equation of McNab (1980b). The lowest temperature at which the animal maintains basal metabolic rate (lower limit of thermoneutrality) was determined following methods in Bonaccorso et al. (1992), i.e., by the intersection of the curve below thermoneutrality (TNZ) with the basal metabolic rate (BMR) and the minimal squares method. We calculate the minimal thermal conductance (C) of the fitted slope of a curve below thermoneutrality; and another that represented an estimate of minimal conductance versus ambient temperature, where we identified the point in which conductance increased. Values for basal metabolic rate and minimal conductance were compared with values expected from standard allometric scaling for mammals and bats: basal metabolic rate (VO_2'/m) for mammals was calculated as mL O_2 g^{-1} h^{-1} = 3.45 $m^{-0.287}$ (where m is the body mass in g; McNab 1988) and standard for bats as mL O_2 g^{-1} h^{-1} = 2.97 $m^{-0.256}$ (Speakman and Thomas 2003). The thermal conductance (C') for mammals was calculated as: mL O_2 g^{-1} h^{-1} $°C^{-1}$ = 1.02 $m^{-0.5}$ (Herreid and Kessel 1967) and for bats as: mL O_2 g^{-1} h^{-1} $°C^{-1}$ = 0.901 $m^{-0.466}$ (Speakman and Thomas 2003). The observed values of VO_2'/m and C' for the species are the percentage of the expected value of the equations allometrics.

Statistical analysis: Numerical values are presented as mean ± 1SE for n number of measurements. N represents the number of individuals. Differences among or between means were tested using a one-way ANOVA or a student's t-test in order to determine differences between physiological variables (Zar 1999). When there were not differences among sexes, the data were joined. Linear regressions were fitted using the method of least squares, and differences between regressions were determined using ANCOVA. Significance was accepted at $P < 0.05$. The statistical package used was JMP 5.0.1.2 (SAS Institute Inc. Campus Drive, Cary, NC, USA, 1989–2003).

8.4 Results

Sturnira bidens: We used 3 males and 11 females in 175 measurements. The mean mass for this species was 17.15 g $\pm$ 0.29 ($N = 14$; range between 14.9 and 18.8 g). The mean mass of the males (17.1 $\pm$ 0.18 g; $n = 23$) was similar to the females (17.5 $\pm$ 0.14 g, $n = 11$) during the experiments ($t = 0.67$; $P = 0.513$). Variations of body temperature, basal metabolic rate, and thermal conductance were not correlated with sex ($F = 0.15$; $P = 0.69$).

S. bidens showed a dichotomous response in Tb to low chamber temperatures. At temperatures below 25 °C, bats either maintained a lightly constant Tb (normothermic state with Tb = 32–34) or dropped Tb below 30 °C to values near Ta (herein defined as torpor; Tb = 19.4–30.8 °C; Fig. 8.1). Both responses were independent of body mass ($t = 1.49$; $P = 0.139$). Normothermic individuals maintained their body temperature (32–34 °C) slightly dependent on the ambient temperature (line a in Fig. 8.1a; Tb = 31.16 + 0.1Ta; the slope of the regression differs significantly from zero, $r^2 = 0.259$; $P = 0.000$; $n = 72$). The mean Tb for all individuals between 19 and 25 °C was 33.5 $\pm$ 0.08 °C ($r^2 = 0.038$; $P = 0.20$; $n = 44$).

Obvious evidence of torpor was observed in individuals exposed to ambient temperatures below 25 °C (line b in Fig. 8.1a; the slope of the regression differs significantly from zero, $r^2 = 0.68$; $P = 0.0001$; $n = 26$). Bats that entered in torpor maintained a minimal body temperature of ca 22 °C. Body temperature fell with decreasing ambient temperature, yet the body temperature of *S. bidens* was somewhat 6 °C above environmental temperature (5.6 $\pm$ 0.41); such variations were not correlated with body mass ($r^2 = 0.08$; $P = 0.17$; $n = 26$). Some individuals exposed to ambient temperatures below 18 °C increased their body temperature at Tb–Ta = 9 °C (white circles in Fig. 8.1a).

As with Tb, metabolic measurements indicated a dichotomous thermoregulatory response (normothermy and torpor) by bats at Ta < 25 °C (Fig. 8.1b). The zone of thermoneutrality extended from 25 to 31 °C (line c in Fig. 8.1b). The mean basal metabolic rate was 1.47 $\pm$ 0.02 mL O_2 g^{-1} h^{-1}, which is equal to 95% of the standard rate expected for mammals and 102% for bats, from a mean mass equal to 16.8 $\pm$ 0.17 g (line c, Fig. 8.1b; the slope of the regression does not differ significantly from zero, $r^2 = 0.002$; $P = 0.72$; $n = 54$). Normothermic metabolism rate increased according to the regression VO_2/m = 5.99–0.18Ta ($r^2 = 0.823$; $P = 0.000$; $n = 72$) at temperatures below thermoneutrality (black circles around line in Fig. 8.1b). Slope of curve of metabolic rate was projected to intercept the axis of ambient temperature at 33.3 °C, with 95% confidence limits covering the range of body temperatures recorded during these experiments. Whereas torpid maintained a variable lower metabolic rate between 0.58 $\pm$ 0.03 and 0.42 $\pm$ 0.02 mL O_2 g^{-1} h^{-1};

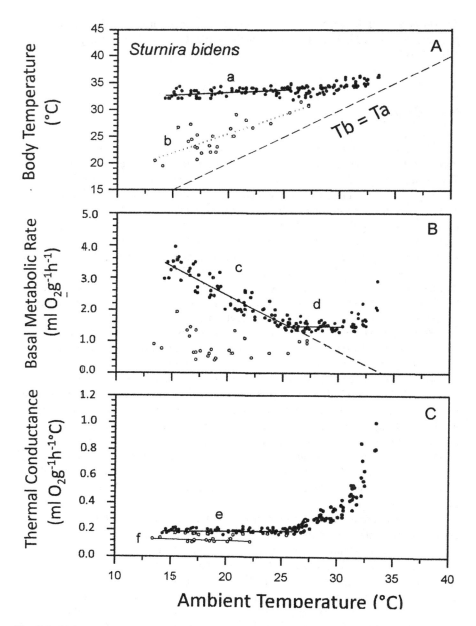

Fig. 8.1 (a) Body temperature, (b) metabolic rate, and (c) humid thermal conductance, as a function of the ambient temperature in *Sturnira bidens* ($N = 14$ individuals). The regression lines are indicated for (*a*) body temperature regulation by normothermics, (*b*) by torpid individuals, (*c*) basal metabolic rate, (*d*) metabolic rate below the thermoneutrality, (*e*) minimal thermal conductance for normothermics, and (*f*) for torpids. The discontinuous line in (a) represents the equality among ambient temperatures (Ta) and body temperature (Tb). The white circles represent the torpid individuals during the experiments

($t = 1.94$; $P = 0.07$; significant heterogeneity of variances, $P = 0.008$). The Tb–Ta differential in thermoconforming torpid bats was not a function of ambient temperature (Fig. 8.1a,b). At low ambient temperature (below 18 °C), torpid bats increased their metabolic rates (1.21 ± 0.2 mL O_2 g^{-1} h^{-1}) to maintain the body temperature above a threshold. Above the upper critical temperature, metabolic rate of all individuals increased, and their body temperature reached 36 °C.

Differences in minimal thermal conductance were found between normothermic and torpid bats ($t = 14.04$; $P = 0.000$; Fig. 8.1c). Minimal thermal conductance of the normothermic bats was independent of the ambient temperature below the 26 °C (line e in Fig. 8.1c; the slope does not differ significantly from zero, $r^2 = 0.013$; $P = 0.31$; $n = 85$), the mean of which is 0.19 ± 0.002 mL O_2 g^{-1} h^{-1} $°C^{-1}$, equal to 79% of the mammalian and 80% of the bats standard, with a body mass of 17.4 ± 0.12 g. In torpid bats, the mean was 0.12 ± 0.004 mL O_2 g^{-1} h^{-1} $°C^{-1}$(line f in Fig. 8.1c), which is 50% of both standards. The slope of regression of the metabolic rate in the lower limit of the thermoneutrality is 95% confidence limits covering the minimal conductance mean (0.18 ± 0.01).

Sturnira bogotensis: Mean body mass of six males and four females ranked from 18.7 to 24.4 g; the mean value was 21.5 ± 0.62 g ($N = 10$) with no differences among sexes ($t = 0.57$; $P = 0.587$). Also, no differences among sexes were found in body temperature, basal metabolic rate, and minimal conductance measurements (ANOVA; $F = 0.42$; $P = 0.47$).

Body temperature of *S. bogotensis* was regulated as normothermic but slightly dependent on ambient temperature (line a in Fig. 8.2a; Tb = 31.42 + 0.054Ta; the slope of the regression differs significantly from zero, $r^2 = 0.104$; $P = 0.017$; $n = 54$). Among Ta = 22 and 25 °C, the Tb was independent of the ambient temperature ($r^2 = 0.085$; $P = 0.2$) with an average of 32.8 ± 0.14 °C ($n = 21$). In 6 of 60 experiments, ambient temperatures remained below 21 °C, and the body temperature fell below 29 °C (circles open in Fig. 8.2a). These individuals were hypothermic and showed lower metabolic rates during the first day of experimentation, yet they gradually recovered normothermy.

The thermoneutrality zone extended from 25 to 31 °C (line b in Fig. 8.2b; the slope did not differ significantly from zero, $r^2 = 0.003$; $P = 0.7$; $n = 47$). Rate metabolic basal average was 1.62 ± 0.02 mL O_2 g^{-1} h^{-1}, or 113% of the standard for mammals and 119% for bats, with a mass of 21.36 ± 0.29 g. At the lower limit of thermoneutrality (25 °C), metabolic rates increased as ambient temperature decreased. According to the regression, VO_2/m = 6.65–0.20Ta (line c in Fig. 8.2b; the slope differed significantly from zero, $r^2 = 0.944$; $P = 0.000$; $n = 54$), and the projection intercepted the axis of ambient temperature at 33.2 °C. The confidence limit of 95% included the body temperature registered for this frugivore during the experiments ($31.75 < $ Tb $ < 33.3$). The individuals with hypothermy showed lower metabolic rates (white circles in Fig. 8.2b; $r^2 = 0.69$; $P = 0.039$; $n = 6$). Above the thermoneutrality, the metabolic rate increased and the animals reached a maximum body temperature of 38 °C.

Minimal thermal conductance differed significantly between normothermic and hypothermic individuals ($t = 9.57$; $P = 0.000$; $n = 58$). The minimal conductance

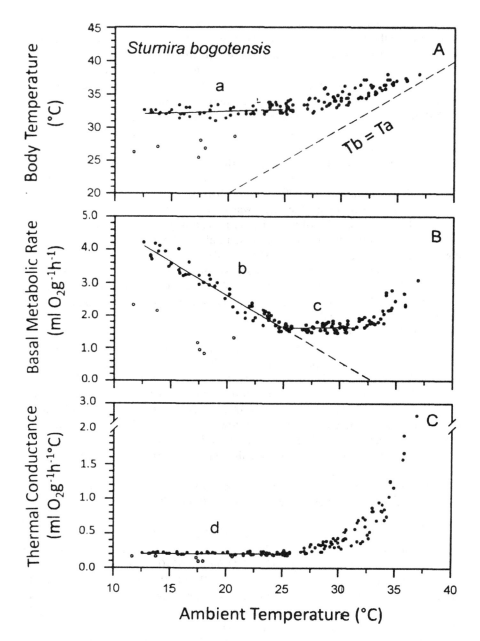

Fig. 8.2 (a) Body temperature, (b) metabolic rate, and (c) humid thermal conductance, as a function of the ambient temperature in *Sturnira bogotensis* ($N = 10$ individuals). The regression lines are indicated for (a) body temperature regulation for normothermics, (b) basal metabolic rate, (c) the metabolic rate below the thermoneutrality, and (d) minimal thermal conductance. The discontinuous line in (**a**) represents the equality between ambient temperatures (Ta) and body temperature (Tb). The white circles represent individual's hypothermics during the experiments

mean was 0.21 ± 0.002 mL O_2 g^{-1} h^{-1} $°C^{-1}$, or 95% of the values expected from a standard for mammals and bats, weighing 21.35 ± 0.28 g (line d in Fig. 8.2c; the slope did not differ significantly from zero, $r^2 = 0.0052$; $P = 0.6051$; $n = 54$; white circles correspond to hypothermic individuals, $n = 6$). The conductance of the slope of regression of the metabolic rate in the lower part of the thermoneutrality zone was inside the interval of confidence of 95%.

Sturnira ludovici: We use four females and seven males in 185 experiments. The mean mass ranked from 23 to 34.3 g, with an average of 28.6 ± 1.17 g ($N = 11$). The mean body mass of males (30.6 ± 1.2 g) was significantly higher than those of females (25 ± 0.89 g; $t = 3.22$; $P = 0.01$); therefore, these data were not combined.

This species showed a normothermic response to environmental temperature changes (Fig. 8.3). Below 25 °C, the females maintained an independent Tb ($r^2 = 0.022$; $P = 0.4015$; $n = 34$), while males show a dependent Tb ($r^2 = 0.113$; $P = 0.024$; $n = 45$). Although we did not find significant differences among sexes (ANOVA; sex, $F = 1.03$; $P = 0.31$), the body temperatures of $S.$ *ludovici* were slightly dependent on the ambient temperature changes (line a in Fig. 8.3a; Tb $= 32.19 + 0.072$Ta, the slope differed significantly from zero, $r^2 = 0.07$; $F = 5.8$; $P = 0.018$; $n = 79$). Mass could explain the differences of thermoregulation; however, at low temperatures, the females were better thermoregulators than males (ANCOVA; mass, $F = 53.2$; $P = 0.000$; sex, $F = 43.67$; $P = 0.0001$; interaction, $F = 0.25$; $P = 0.62$). At ambient temperatures ranging between 18 and 25 °C, the mean body temperature was 33.8 ± 0.13 °C ($n = 43$). In six experiments, some individuals entered in hypothermy at ambient temperatures below 21 °C, with a variable body temperature between 21 and 30 °C.

The thermoneutral zone extended to 32 °C (Fig. 8.3b). We did not find differences among sexes concerning basal metabolic rates (ANCOVA; mass, $F = 34.59$; $P = 0.000$; sex $F = 0.17$; $P = 0.67$; interaction, $F = 1.48$; $P = 0.23$; line b in Fig. 8.3b). The average of the basal metabolic rate was 1.74 ± 0.02 mL O_2 g^{-1} h^{-1} ($n = 82$), or 131% and 137% of the value standard for mammals and bats, respectively, with a mass average of 27.4 ± 0.4 g. Below thermoneutrality (25.5 °C), the metabolic rate varied with sex ($F = 46.98$; $P = 0.000$), which we can explain by body mass differences (ANCOVA; mass, $F = 56.48$; $P = 0.000$; sex, $F = 1.15$; $P = 0.287$; interaction, $F = 2.67$; $P = 0.11$). Therefore, we analyzed the values jointly, and we obtained the following regression: $VO_2/m = 7.273–0.216$Ta (line c in Fig. 8.3b; the slope differed significantly from zero, $r^2 = 0.824$; $P = 0.000$; $n = 79$). The slope of the line of regression of the metabolic rate intercepted the axis of ambient temperature at 33.7 °C and the limits of confidence included body temperature values registered during the experiments ($33.4 <$ Tb < 33.8). A few hypothermic individuals maintained lower metabolic rates (white circles in Fig. 8.3b). Above the thermoneutral zone, the maximum body temperature was 39.4 °C.

We did not find differences among sexes for minimal thermal conductance (ANCOVA; mass, $F = 25.85$; $P = 0.0001$; sex, $F = 2.6$; $P = 0.11$; interaction, $F = 2.49$; $P = 0.12$). Minimal conductance determined the lower limit of thermoneutrality (line d in Fig. 8.3c; the slope of the regression did not differ significantly

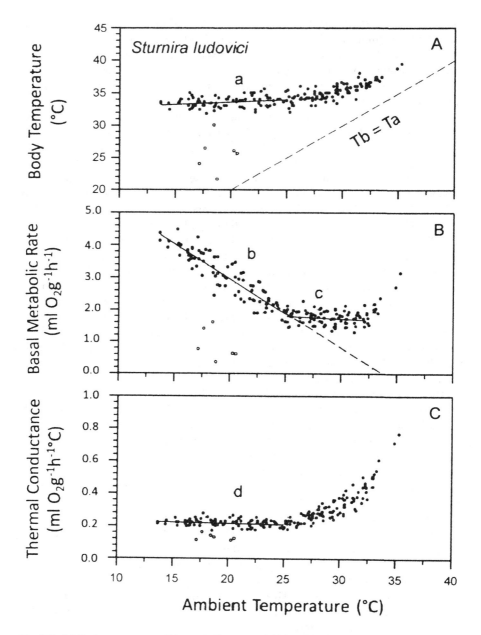

Fig. 8.3 (a) Body temperature, (b) metabolic rate, and (c) humid thermal conductance, as a function of the ambient temperature in *Sturnira ludovici* ($N = 11$ individuals). The regression lines are indicated for (*a*) body temperature regulation by normothermics, (*b*) basal metabolic rate, (*c*) the metabolic rate below the thermoneutrality, and (*d*) minimal thermal conductance. The discontinuous line in (*a*) represents the equality between ambient temperatures (Ta) and body temperature (Tb). The white circles represent individual's hypothermics during the experiments

from zero, $r^2 = 0.034$; $P = 0.101$; $n = 79$). The mean minimal conductance was 0.22 ± 0.002 mL O_2 g^{-1} h^{-1} °C^{-1}, or 116% of the value expected for mammals and bats, with a mass of 29.5 ± 0.5 g. The mean minimal conductance was inside the interval of confidence of 95%, of the slope of regression of metabolic rate (0.216 ± 0.01). The hypothermic individuals (white circles in Fig. 8.3c) showed a mean conductance (0.13 ± 0.007) significantly lowest ($t = 9.6$; $P = 0.000$; $n = 6$).

Comparison among species: Body temperature varied significantly among the species (ANOVA; $F = 15.52$; $P = 0.000$). These differences may be related to body mass (mass, $F = 584.97$; $P = 0.000$; *S. bidens* < *S. bogotensis* < *S. ludovici*). We found that when ambient temperatures are low, differences in body temperature are independent of body mass (ANCOVA; mass, $F = 1.14$; $P = 0.29$; species, $F = 7.15$; $P = 0.001$; Ta, $F = 0.004$; $P = 0.95$; interaction, $F = 2.17$; $P = 0.12$). An a posteriori test of multiple comparisons (Tukey HSD) showed that *S. bidens* and *S. bogotensis* maintain significantly lower body temperatures than *S. ludovici*.

The metabolic rate in the thermoneutral zone varied significantly among species (ANCOVA; mass, $F = 143.84$; $P = 0.0001$; species, $F = 3.63$; $P = 0.022$; Ta, $F = 0.21$; $P = 0.64$; interaction, $F = 0.30$; $P = 0.76$); only *S. ludovici* showed a significantly higher basal rate than that was expected for its mass (Tukey HSD). The minimal conductance was significantly different among them (ANCOVA; mass, $F = 166.04$; $P = 0.0001$; species, $F = 8.50$; $P = 0.0003$; Ta, $F = 1.57$; $P = 0.21$; interaction, $F = 0.51$; $P = 0.60$). The test of Tukey HSD showed that *S. bogotensis* and *S. ludovici* had higher minimal conductance. The length of the fur was significantly different among the three species ($F = 13.23$; $P = 0.0001$; Tukey HSD), yet it failed to explain the differences. The individuals of *S. ludovici* had shortest coat (7.59 ± 0.09; $N = 21$) than those of *S. bidens* (8.05 ± 0.1; $N = 20$) and *S. bogotensis* (8.3 ± 0.11; $N = 20$).

8.5 Discussion

Thermoregulation: *Sturnira ludovici* and *S. bogotensis* were normothermics below thermoneutrality, and they maintained a high and constant body temperature over a wide range of air temperatures. However, their body temperatures were lower than those registered for other lowland fruit bats (ANOVA; $F = 7.45$; $P = 0.02$; $N = 16$; Table 8.1). In contrast, *S. bidens* showed a dichotomous response (heterothermy) to low air temperatures, one normothermic and the other defined as daily torpor. Indeed, only one individual of *S. bidens* showed both responses, but it was not possible to establish a clear threshold of metabolism between normothermy and torpor. According to these results, neither selective sex-biased nor body mass affect the faculty of entering torpor. These differences among individuals of *S. bidens* suggest that torpor stage is not an obligate response to lower air temperatures. The minimum body temperature of *S. bidens* reported here (ca. 22 °C) is similar to those registered in torpid tropical nectarivorous and insectivorous bats of equivalent body mass (Bonaccorso and McNab 1997; Genoud and Bonaccorso 1986). Likewise during torpor, *S. bidens* failed to reduce its body temperature below 22 °C, which could

Table 8.1 Energetic comparison among frugivore bats from families Phyllostomidae and Pteropodidae

Species	Body mass (g)	BMR	%BMR	C	%C	F	Tb	Tcl	Tcm	TS	Shelter	D	E	R	T	TR
Phyllostomidae																
Artibeus concolor[1]	19.7	1.67	113.87	0.21	90.95	1.3	35.3	28			Foliage	C	L	TR	N	G
A. fimbriatus[2]	63.9	1.22	116.61								Foliage	C	L	TR	N	?
A. jamaincensis[1]	45.2	1.25	108.17	0.14	92.28	1.2	36.4	25	35.8		Foliage	C	L	TR	N	G
A. lituratus[1,2]	69.7	1.24	121.51	0.11	90.03	1.3	37.3	25	37.7	19	Foliage/caves	C	L	TR	N	G
Carollia perspicillata[1,2]	13.9	2.10	129.55	0.27	98.69	1.3	36.4	28.2	37.7	24.3	Caves/hollow trees	C	L	TR	N	G
Chiroderma doriae[2]	19.9	1.56	106.68							26.4	?	C	L	TR	N	?
Platyrrhinus lineatus[1,2]	22.3	1.38	97.50	0.19	88.89	1.1	36.4	28			Foliage	C	L	TR	N	G
Rhinophylla pumilio[1]	9.5	1.71	94.58	0.31	93.37	1.0	34.7	30	37		?	C	L	TR	N	G
Sturnira bidens[3]	16.8	1.47	95.76	0.19	76.35	1.3	33.5	25	33	15	Hollow trees?	C	H	TR	Y	P
S. bogotensis[3]	21.4	1.62	113.12	0.21	95.24	1.2	32.8	25	37	15	Hollow trees?	C	H	TR	N	I
S. erythromos[4]	15.9	2.01	128.88	0.26	101.64	1.3	34.4	25.5	36	15	Hollow trees?	C	H	ST	Y	P
S. lilium[1]	21.9	1.79	125.82	0.19	87.17	1.4	36.4	28.1	37.1	28.4	Hollow trees	C	L	ST	N	G
S. ludovici[3]	27.4	1.74	130.43	0.22	112.90	1.2	33.8	26	35	15	Hollow trees?	C	A	TR	N	I
S. tildae[2]	20.5	1.95	134.49								Hollow trees?	C	L	TR	N	?
Uroderma bilobatum[1]	16.2	1.64	105.72	0.25	98.65	1.1	35.1	28	35.6		Foliage	C	L	TR	N	G
Vampyressa pussila[1]	8.8	2.11	114.16							27.3	?	C	L	TR	N	?
													L			
Pteropodidae													L			
Cynopterus brachyotis[5]	37.4	1.27	104.09	0.19	113.92	0.9	36.5	30	36		Trees	C	L	TR	N	G
Dobsonia anderseni[6]	241.4	0.72	100.78	0.09	131.00	0.8	36.4	28	34		Caves	I	L		N	G
D. minor[6]	73.7	1.01	100.57	0.12	101.00	1.0	36.1	28	35		Trees	C	L	TR	N	G
D. moluccensis[6]	404.3	0.91	147.69	0.09	177.42	0.8	36.8	27	36		Caves	C	L	TR	N	G
D. praedatrix[6]	179.5	0.79	102.07	0.08	105.08	1.0	37.1	27	32.5		Caves	I	L	TR	N	G
Nyctimene albiventer[6]	30.9	0.88	68.28	0.09	50.68	1.3	35.9	28	33.5		Trees	C	B	TR	N	I

8 Adaptive Strategies of Frugivore Bats to Andean Cloud Forests

Species	Body mass (g)	BMR	%BMR	C	%C	F	Tb	Tcl	Tcm	TS	Shelter	D	E	R	T	TR
N. cyclotis[6]	40.4	1.60	134.07	0.09	53.59	2.5	36	17	34		Trees	C	A		N	G
Paranyctimene raptor[6]	23.6	1.04	74.69	0.15	71.44	1.0	33.8	27	35		Trees	C	B	TR	Y	I
Pteropus giganteus[7]	(562.2)	0.52	92.77	0.02	46.49	2.0	36.7	30			Trees	C	B	TR	N	G
P. hypomelanus[7]	(520.8)	0.56	97.74	0.03	58.17	1.7	35.7	30			Trees	I	B	TR	N	G
P. poliocephalus[8]	598.0	0.53	96.24	0.02	55.14	1.7	37	17	35		Trees	C	B	ST	N	G
P. pumilus[7]	194.2	0.65	85.47	0.05	68.31	1.3	36.1	23			Trees	I	B	TR	N	G
P. rodricensis[7]	254.5	0.53	75.32	0.05	78.20	1.0	36.5	24			Trees	I	B	TR	N	G
P. vampyrus[7]	1024.3	0.78	165.30	0.03	103.54	1.6	36.9	30			Trees	C	B		N	G
Rousettus aegyptiacus[9]	146.0	0.84	101.77	0.10	118.46	0.9	34.8	31	34		Trees	C	B	ST	N	G
R. amplexicaudatus[6]	91.5	1.14	120.79	0.11	103.16	1.2	36.5	26	34		Caves	C	B	TR	N	G

Source of data: [1]McNab (1982a, b), [2]Cruz-Neto et al.(2001), [3]This study, [4]Soriano et al. (2002), [5]McNab (1989), [6]McNab and Bonaccorso (2001), [7]McNab and Armstrong (2001), [8]Bartholomew et al. (1964), [9]Noll (1979)

Abbreviations: BMR mass specific basal metabolic rate; C mass-specific thermal conductance; $\%BMR$ percent of spected value of basal metabolic rate (McNab 1988), $\%C$ spected value of thermal conductance (Herreid & Kessel 1967), F quotient between spected value of basal metabolic rate and thermal conductance, Tb body temperature, Tcl lower critical temperature, Tcm maximal critical temperature, TS temperature of shelter, D distribution, C continental, I insular, E elevation, L lowlands, H highlands, R region, TR tropics, ST subtropics, T torpor, N not enter in torpor, Y enter in torpor, TR regulation of body temperature, G good, I intermediate, P poor,? data not available. For species that showed several values in bibliography for a parameter done, average was taken if difference was ≤5% or the lower value if that difference was >5%. For species with sexual dimorphism, average values were taken and represented in brackets represent a limitation in its capacity to occupy temperate zones (Bonaccorso and McNab 1997; Geiser et al. 1996; Law 1994).

The capacity of entering torpor has not hitherto been demonstrated in the frugivorous phyllostomid bats. Our results suggest that it occurs on a daily basis in *S. bidens*, as well as *S. erythromos* (Soriano et al. 2002). Other authors such as Audet and Thomas (1997), Cruz-Neto and Abe (1997), and Studier and Wilson (1970) have demonstrated that the phyllostomids can be poor thermoregulators in the laboratory and also under natural conditions, as a consequence of deficient nutritional state. We could explain the hypothermy during the experiments in all of three species studied here, as a consequence of absence of brown fat, as well as a poor nutrition before their capture (A. Ruiz, pers. obs.). As we fed them ad libitum and some individuals were able to recover during the period of captivity (three or four days), this allowed us to determine the true use of torpor in the individuals of *S. bidens* that recovered their normothermic body temperature when experiments concluded.

In tropical highland species, we observed a reduction in the lower critical limit of thermoneutrality (to 25 °C), which could represent a decrease in thermoregulatory cost, similar to the decrease observed in birds of temperate zones (Canterbury 2002). Therefore, we correlated the decrease of their lower critical temperature with altitude, when body mass was deleted from the analysis (ANCOVA; mass, $F = 48.13$; $P = 0.0001$; altitude, $F = 37.39$; $P = 0.0001$). Low air temperature thermoregulation is mass-independent substantially reduced energy expenditure. If we compare the temperature differentials in thermoneutrality ($\Delta T = T_{body} - T_{limit}$) of our species of *Sturnira* with some of other genera (i.e., small *Artibeus*, *Rhinophylla*, *Vampyressa*; Table 8.1), it seems to be that *Sturnira* and other small frugivores that reach marginally or that inhabit the mountain ecosystems (such as *Carollia*) can maintain similar temperature differentials than those recorded for lowland species. The differences among these species were observed at maximum ambient temperatures (Table 8.1); therefore, animals exposed to higher temperatures (>33 °C) were not able to maintain their body temperature below the ambient temperature for long periods. At high ambient temperatures, these bats depend heavily on evaporative cooling to keep their body temperature below lethal levels (Bartholomew et al. 1970).

For phyllostomid bats, it has been demonstrated that thermoregulation is influenced by body mass (McNab 1969, 1970, 1983). In these group of bats, body temperatures were variable (34.9 ± 0.33) and clearly correlated with their body mass ($F = 10.75$, $P = 0.008$, $r^2 = 0.73$; $N = 16$). When we deleted body mass from the analysis, the elevation appeared to be significant (ANCOVA; $F = 15.96$; $P = 0.0025$; Test of Tukey HSD), yet basal metabolic rates were not ($F = 0.89$; $P = 0.36$; $r^2 = 0.38$; $N = 16$). As mentioned earlier, we can interpret that a frugivorous diet does not imply that these results enable us to affirm that regulation of the body temperature depends directly on the intake of high-energy food. In these species, temperature regulation appears to be related to body size rather than diet and could be dependent on the maximum adjustments of their metabolic rates and thermal conductance.

Rate of metabolism: In this study, we were not able to demonstrate a higher BMR for all of bat species from highlands (Table 8.1). Only, *S. ludovici* showed

BMR that exceeds those of lowland fruit bats (137%) with equivalent body mass. Some authors have suggested that higher basal metabolic rates in *Sturnira erythromos* are an adaptation to highland environments (McNab 2003; Soriano et al. 2002). But the short time of experimentation used by Soriano et al. (2002) for *S. erythromos* (~1.5 h) suggests that average values calculated correspond to the rest metabolic rate (2.51 mL O_2 g^{-1} h^{-1}). Consequently, if the rest metabolic rate is equal to 1.25 times the BMR (Aschoff and Pohl 1970; Grodzinski and Wunder 1975; Kendeigh 1970), then BMR of *S. erythromos* should be 2.01 mL O_2 g^{-1} h^{-1} or 129% (m = 15.9 g). If we compare its BMR with those observed in *S. bidens* (96%; m = 16.8 g), we can obtain a 30% difference among both species, which related the maximum thermogenic capacity of small fruit bats (Hayes and Garland 1995).

The metabolic rates of daily torpor (TMR) for *Sturnira bidens* and *S. erythromos* (0.42 and 0.63 mL O_2 g^{-1} h^{-1}, respectively) lead to a significant greater energy saving (between 69% and 71%, respectively). At Ta = 17–18 °C, we calculated a Q_{10} between the basal metabolic rate (BMR) and torpor rate (TMT) of 2.68 for *S. bidens* and of 2.1 for *S. erythromos*, at body temperature of 22 °C, values that are typical for biochemical reactions (Q_{10} = 2–3; Schmidt-Nielsen 1990). Thus, temperature dependence and not physiological inhibition seems to be the factor involved in lowering the metabolic rate of *S. bidens* and *S. erythromos*, compared with small hibernating species (Bartels et al. 1998; Geiser 1988; Hosken and Withers 1999). The body temperature of these torpid bats seems to be related to ambient temperature during torpor (17–18 °C). In Andean cloud forests, the minimum air temperature remains below 20 °C, which resulted in energetic savings, but at Ta < 18 °C, Tb and metabolic rates of *S. bidens* were more variable and tended to increase, resulting in smaller energy savings (Fig. 8.1). Therefore, both species will be able to live in environments in which ambient temperatures do not decrease below the torpor temperature and where the highest energy saving is given. The use of adaptive heterothermy in the smallest species of fruit bats is not related with BMRs, which fall both above and below the boundary line of endothermy (Cooper and Geiser 2008; McNab 1983).

It is very hard to predict the impact of food habits on basal metabolic rate, since the specific food habits of some species are poorly known, and they are both geographic and seasonally variable (McNab 1986,2003). In the Venezuelan cloud forests, these sympatric species feed strictly on *Piper*, *Vismia*, and *Solanum* species and Araceae family (Ruiz 2006); the higher BMRs of fruit-eater bats have been associated with the consumption of highly nutritional fruits (McNab 2003). For example, the high metabolic rates of *S. bogotensis* and *S. ludovici* may be associated with feeding on fruits of *Solanum* and *Piper*, respectively (Ruiz 2006). In contrast, both *S. erythromos* and *S. bidens*, which feed on *Solanum* and Araceae species, both use torpor (Ruiz 2006). Therefore, the quality of the fruits is not the only factor responsible for these differences (Dinerstein 1986; Herbst 1983 cited by Fleming 1988; Herbst 1985). However, if fruits are scarce during certain periods (Fleming et al. 1993; McNab 2003; Morrison 1980; Thomas 1984), one possible response would be to decrease the energy expenditure (i.e., decrease BMR and/or enter in torpor). The fruits of *Solanum* and *Piper* are available all year round, while Araceae fruits are seasonal (Soriano 1983). It is possible that the endothermy in these species will be associated with resource availability; otherwise, they would enter torpor (Delorme and Thomas 1996,1999). Indeed, differences in body size between these three species and also *S. erythromos* (Soriano et al. 2002) could explain the fast depletion of energy stores of species with body mass lower than 17 g (such as *S. bidens* and *S. erythromos*), whereas those with body mass larger than 20 g (such as *S. bogotensis* and *S. ludovici*) are able to maintain normothermy for long time.

One of the physiological consequences of a frugivorous diet could be a higher digestive efficiency to regulate fruit intake in relation to protein intake (Delorme and Thomas 1996,1999; Dumont et al. 1999; Herrera et al. 2002; Korine et al. 1999; Schondube and Martínez del Rio 2004;Thomas 1984). Many birds and mammals increase food ingestion rates when acclimated to cold temperatures (Hammond and Diamond 1997). However, we do not have any evidence that such changes in the digestive function are operating in *Sturnira*.

The dependence of basal metabolic rates on body mass can be observed in all Neotropical fruit bats ($F = 134.6$; $P = 0.0001$; $r^2 = 0.91$; $N = 16$). This relationship described by the equation: VO_2 (mL O_2 h^{-1}) $= 3.66$ m$^{0.74}$, whose slope slightly differs from that obtained by McNab (1988) for all the mammals (0.713). Aside from body mass, other factors such as phylogeny, foraging strategy, food preferences, and roosting sites, have been used to explain the variations of BMR in phyllostomid bats (Bonaccorso and McNab 2003; Cruz-Neto and Bozinovic 2004; Cruz-Neto et al. 2001; McNab 2003; McNab and Bonaccorso 2001). We used altitude, torpor, and taxonomic affinities, yet their basal metabolic rate was not correlated with any of these factors (ANCOVA; $F = 0.31$; $P = 0.59$). Other factors proposed by McNab (2003) were not analyzed, since we did not consider that they provided accurate data to carry out intra- and inter-specific comparisons. Our data demonstrated that members of genus *Sturnira* seem to evolve different adaptations that allowed them to face a single environmental factor (such as ambient temperature). An examination on the composition of their diets, availability, seasonality of resources, and roost microclimate will be required to explain these differences (i.e., growth and reproduction).

Thermal conductance. The normothermic species studied showed minimal conductances similar to lowland fruit-eater bats (ANCOVA; mass, $F = 7.31$; $P = 0.022$; altitude, $F = 0.173$; $P = 0.67$; $N = 13$; Table 8.1). The length of the fur could explain partiality in the variations observed among *Sturnira* spp. For example, the individuals of *S. ludovici* exhibited the shortest fur, coinciding with a highest minimal conductance (112%), while *S. bidens* and *S. bogotensis* showed the longest fur with a lowest conductance (79 and 95%, respectively). Torpid *S. bidens* exhibited lower conductance values than normothermic individuals. Lower conductance during torpor (50% of expected) has been observed in insectivorous bat species (Cryan and Wolf 2003; Genoud 1993; Hosken and Withers 1997,1999). It is unclear why conductance is sometimes lower during torpor, but possible explanations include changes in breathing rates, posture, and circulation (Hosken and Withers 1997,1999).

8 Adaptive Strategies of Frugivore Bats to Andean Cloud Forests 169

The roosting sites used by Andean *Sturnira* species are not well known. Until the present, it has been reported that at altitudes of 1800 m, *S. bidens* and *S. aratathomasi* possibly use caves (Tamsitt et al. 1986), while lowland forest dwellers such as *S. lilium* live in hollow trunks, palms, or lianas (Fenton et al. 2000). In the study area, we found individuals of *Carollia brevicauda* living in hollow trunks (A. Ruiz, pers. obs.). Therefore, we consider probable that *Sturnira* also is using these kinds of roosts. We have estimated that average air temperature values inside these eventual roosts are 15.4 °C ± 0.4 (A. Ruiz, pers. obs.). Nevertheless, this value is below the lower limit of thermoneutrality. It is probable that gregarious behavior observed in captivity is also present under natural conditions as a mechanism that allows them to reduce heat loss, mainly in smaller species (Ashton et al. 2000; Mayr 1963; McNab 1971; Studier 1970; Thomas and Cloutier 1992).

The influence of body mass on minimal conductance in Neotropical frugivores is described by the equation $C = 1.72$ m$^{0.324}$, where 42% of the variation in conductance is explained by their mass (ANCOVA; $F = 7.94$; $P = 0.017$; $N = 13$; Table 8.1). Most of these fruit-eater bats showed conductance values below that was expected for mammals and were not associated to the basal metabolic rates (ANCOVA; $F = 0.88$; $P = 0.37$; $N = 13$), roosts (ANCOVA; $F = 0.47$; $P = 0.79$; $N = 13$; Table 8.1), and ambient temperature (ANCOVA; $F = 4.12$; $P = 0.21$). The lowest conductance observed in small fruit-eater bats (<50 g) of high- and lowlands could explain their ability to maintain their body temperature above ambient temperature without making adjustments in their metabolic rates. However, in torpid individuals it was not enough to maintain the normothermy, as a consequence of their high surface/volume ratio values.

Altitudinal distribution and physiological limits: *Sturnira erythromos* and *S. ludovici* occur throughout Neotropical mountains and penetrate subtropical latitudes (Gannon et al. 1989; Giannini and Barquez 2003; Simmons 2005). In the northern Andes of Argentina, *S. erythromos* is sympatric with another species of the genus, such as *S. lilium*, which was later restricted to lowlands (<1200 m), while the former replaced it at higher elevations (Giannini 1999). Seasonality and distribution of resources in subtropical latitudes should explain the altitudinal segregation of both species (Giannini 1999); however, if they are eating similar fruits, why are they primarily restricted to tropical highlands and lowlands, respectively? Ecological or historical factors may be utilized for answering it, but the similarities in their diet would be correlated with their physiology. Nevertheless, both species with high BMR (between 126% and 130% of expected) show differences when the temperature is below the lower critical temperature. The use of torpor at lower temperatures by *S. erythromos* may be a physiological trait that has been evolved mass dependent in some highland fruit-eater bats.

However, *S. ludovici* showed metabolic rates above those registered in lowland fruigivorous bats; in fact, it was the only of the three studied species with a broad altitudinal distribution (500–3000 m). Although *S. ludovici* can be considered a complex species (Timm and LaVal 1998), in the Andes of Venezuela this species is below 2200 m of elevation, yet we have captured at higher elevations (>2000 m). Its intermediate size could facilitate altitudinal movements in search of resources that may satisfy their high-energy and nutritional requirements. In contrast, although *S. bogotensis* and *S. bidens* are restricted mainly to tropical highlands (>2000 m; Molinari and Soriano 1987; Simmons 2005), their thermoregulatory patterns differ in relation to their body mass. The absence of daily torpor in *S. bogotensis* could be related with its capacity for maintaining high BMR, thus avoiding heat loss due to a smaller relationship surface to volume ratio (Table 8.1).

The ranges of altitudinal distribution of the *Sturnira* spp. are wide, and few species are exclusive of the highlands (>2000 m). There are contradictory hypotheses concerning the evolutionary history and ancestral origin of genus *Sturnira: one ancestor of lowland or one ancestor of highland* (de la Torre 1961; Iudica 2000; Koopman 1982; Owen 1990; Pacheco and Patterson 1991; Villalobos and Valerio 2002). If this genus evolved in the mountain environments, the torpor must be a primitive character, because it first appeared in *S. bidens*, which is a member of basal group (Pacheco and Patterson 1991; Villalobos and Valerio 2002). However, the torpor in genus *Sturnira* seems to have evolved twice, once within the clade that belongs to *S. bidens* (coming from a highland ancestor) and again in the clade that belongs to *S. erythromos* (coming from a lowland ancestor). Herein, phylogeny and food habits cannot be the explanation since *S. bogotensis* (a brother taxon of *S. erythromos*) does not use the torpor (if this were adaptive), under the assumption that they are living under similar environmental pressures. After all, a species is unlikely to maintain characteristics that are incompatible with a larger size and higher BMR.

In addition, both adaptation to low temperatures and restricted altitudinal distribution are not explained by the heterothermy. There are several evidence that show that the torpor is not associated to the altitude, as we did think. Some studies of fruit-eater bats of the old world (Pteropodidae) that inhabit in lowland show a similar pattern with the one observed in *Sturnira*, in which smaller species (23-40 g) are also poor thermoregulators and torpor is infrequent. In contrast, intermediate and larger species (73–1030 g) reach montane ecosystems due to their appropriate thermoregulatory capacity (Bartels et al.1998; Bonaccorso and McNab 1997; Law 1994; McNab and Armstrong 2001; McNab and Bonaccorso 2001). The choice of a body size limit for small fruit bats that enter or not in torpor requires more objectivity. The recognition of small and large bat body size classes has important implications in metabolic rates, the evolution of life histories, and ultimately, fitness in these species (Lovegrove 2005).

The physiologic responses of these species to face environmental temperature fluctuations depend on their body size. The dichotomous thermoregulatory responses used by the smallest species with the genus *Sturnira* as well as the normothermy observed in the largest ones suggest that its fruit-based diet does not compensate the energy cost of the thermoregulation at highlands for all of them. Therefore, the ultimate factor that explains the thermoregulatory patterns used by these species is their body mass. We suggest that the adaptation of these species to Andean highlands depends on their thermoregulatory capacity, conditioned by their body size and, probably, by the availability and quality of fruits that they consume. However, the apparent physiologic inability to reduce the metabolic rates and their body temperature when they are exposed to low temperatures could explain their restricted distribution to the tropical highlands.

Acknowledgments We thank A. Arends and C. Bosque who provided useful comments about an earlier version of this manuscript, and G. Alba y C. Santiago helped us with the equipment and software installation. J. Murillo, M. Machado, and C. Cabrera and all the students of Animal Ecology Laboratory provided field and laboratory assistance. F. Ely helped us with the English revision. We received partial financial support through the grant from Red Latinoamericana de Botánica (RLB-Chile), Scott Neotropical Fund of Cleveland Metroparks Zoo, Latin American Fellowship of the American Society of Mammalogist, Consejo de Estudios Científicos y Humanísticos (ULA, Venezuela; Project No. C-1097-01-01-ED), Postgrados Integrados en Ecología-FONACYT (Venezuela), Idea Wild Assistance Support, and COLCIENCIAS (Colombia).

References

Alberico M, Cadena A, Hernández-Camacho J, Muñoz-Saba Y (2000) Mamíferos (Synapsida: Theria) de Colombia. Biota Colombiana 1(1):43–75

Aschoff J, Pohl H (1970) Der Ruheumsatz von Vögeln als Funktion der Tageszeit und Körpergröβe. J Ornithol 111:38–47

Ashton KG, Tracy MC, Queiroz A (2000) Is Bergmann's rule valid for mammals? Am Nat 156(4):390–415

Audet D, Thomas DW (1997) Facultative hypothermia as a thermoregulatory strategy in the phyllostomid bats, *Carollia perspicillata* and *Sturnira lilium*. J Comp Physiol B 167:146–152

Bartels W, Law BS, Geiser F (1998) Daily torpor and energetic in a tropical mammal, the northern blossom-bat *Macroglossus minimus* (Megachiroptera). J Comp Physiol B 168:233–239

Bartholomew GA, Dawson WR, Lasiewski RC (1970) Thermoregulation and heterothermy in some of smaller flying foxes (Megachiroptera) of New Guinea. J Comp Physiol 70:391–404

Bonaccorso FJ, McNab BK (1997) Plasticity of energetic in blossom bats (Pteropodidae): impact on distribution. J Mammal 78(4):1073–1088

Bonaccorso FJ, McNab BK (2003) Standard energetic of leaf-nosed bats (Hipposideridae): its relationship to intermittent- and protracted-foraging tactics in bats and birds. J Comp Physiol B 173:43–53

Bonaccorso FJ, Arends M, Genoud D, Cantoni D, Morton T (1992) Thermal ecology of moustached and ghost-faced bats (Mormoopidae) in Venezuela. J Mammal 73(2):365–378

Canterbury G (2002) Metabolic adaptation and climatic constraints on winter bird distribution. Ecology 83(4):946–957

Coburn DK, Geiser F (1998) Seasonal changes in energetic and torpor patterns in the subtropical blossom-bat *Syconycteris australis* (Megachiroptera). Oecologia (Berlin) 113:467–473

Colwell RK, Lees DC (2000) The mid-domain effect: geometric constrains on the geographic of species richness. Trends Ecol Evol 15:70–76

Contreras-Vega M, Cadena A (2000) Una nueva especie del género *Sturnira* (Chiroptera: Phyllostomidae) de los Andes colombianos. Revista de la Academia Colombiana de CienciasXXIV 91:285–287

Cooper CE, Geiser F (2008) The "minimal boundary curve for endothermy" as a predictor of heterothermy in mammals and birds. Rev J Comparative Physiol 178:1–8

Cruz-Neto AP, Abe AS (1997) Taxa metabólica e termorregulaçâo no morcego nectarívoro, *Glossophaga soricina* (Chiroptera, Phyllostomidae). Rev Bras Biol 57(2):203–209

Cruz-Neto AP, Bozinovic F (2004) The relationship between diet quality and basal metabolic rate in endoterms: insights from intraspecific analysis. Physiol Biochem Zool 77(6):877–889

Cruz-Neto AP, Garland TJ, Abe AS (2001) Diet, phylogeny, and basal metabolic rate in phyllostomid bats. Fortschr Zool 104:49–58

Cryan PM, Wolf BO (2003) Sex differences in the thermoregulation and evaporative water loss of a heterothermic bat, *Lasiurus cinereus*, during its spring migration. J Exp Biol 206:3381–3390

De laTorreL (1961) The evolution variation and systematic of the Neotropical bats of the genus Sturnira. Dissertation, University of Ilinois, Urbana

Delorme M, Thomas DW (1996) Nitrogen and energy requirements of the short-tailed fruit bat (*Carollia perspicillata*): fruit bats are not nitrogen constrained. J Comp Physiol B 166:427–434

Delorme M, Thomas DW (1999) Comparative analysis of the digestive efficiency and nitrogen and energy requirements of the phyllostomid fruit-bat (*Artibeus jamaicensis*) and the pteropodid fruit-bat (*Rousettus aegyptiacus*). J Comp Physiol B 169:123–132

Dépocas F, Hart JS (1957) Use of the Pauling oxygen analyzer for measurement of oxygen consumption of animals in open- circuit systems and in a short-lag, closed-circuit apparatus. J Appl Physiol 10(3):388–392

Dinerstein E (1986) Reproductive ecology of fruit bats and the seasonality of fruit production in a Costa Rican cloud forest. Biotropica 18(4):307–318

Dumont ER, Etzel K, Hempel J (1999) Bat salivary proteins segregate according to diet. Mammalia 63:159–166

Fenton MB, Vonhof MJ, Bouchard S, Gill SA, Johnston DS, Reid FA, Riskin DK, Standing KL, Taylor JR, Wagner R (2000) Roots used by *Sturnira lilium* (Chiroptera: Phyllostomidae) in Belize. Biotropica 32(4a):729–733

Fleming TH (1988) The short-tailed fruit bat.A study in plant-animal interactions. The University of Chicago Press, Chicago

Fleming TH, Nuñez RA, Sternberg L (1993) Seasonal changes in diets of migrant and non-migrant nectarivorous bats as revealed by carbon stable isotope analysis. Oecologia (Berlin) 94:72–75

Gannon MR, Willig MR, Knox Jones JJ (1989) Sturnira lilium. Mamm Species 333:1–5

Gannon WL, Sikes RS, The Animal Care and Use Committee of the American Society of Mammalogists (2007) Guidelines of the American Society of Mammalogists for the use of wild mammals in research. J Mammal 88:809–823

Geiser F (1988) Reduction of metabolism during hibernation and daily torpor in mammals and birds: temperature effect or physiological inhibition? J Comp Physiol B 158:25–37

Geiser F, Coburn DK (1999) Field metabolic rates and water uptake in the blossom-bat *Syconycteris australis* (Megachiroptera). J Comp Physiol B 169:133–138

Geiser F, Coburn DK, Körtner G (1996) Thermoregulation, energy metabolism, and torpor in blossom-bats, *Syconycteris australis* (Megachiroptera). J Zool 239:583–590

Genoud M (1993) Temperature regulation in subtropical tree bats. Comparative Biochem Physiol 104A:321–331

Genoud M, Bonaccorso FJ (1986) Temperature regulation rate of metabolism, and roost temperature in the greater white-lined bat *Saccopteryx bilineata*(Emballonuridae). Physiol Zool 59(1):49–54

Giannini NP (1999) Selection of diet and elevation by sympatric species of *Sturnira* in an Andean rainforest. J Mammal 80(4):1186–1195

Giannini NP, Barquez RM (2003) Sturnira erythromos. Mamm Species 729:1–5

Graham GL (1983) Changes in bat species diversity along an elevational gradient up the peruvian Andes. J Mammal 64(4):559–571

Graham GL (1990) Bats versus birds: comparisons among Peruvian volant vertebrates faunas along an elevational gradient. J Biogeogr 17:657–668

Grodzinski W, Wunder BA (1975) Ecological energetic of small mammals. In: Golley FB, Petrusewics K, Ryszkowiski L (eds) Small mammals, their productivity and population dynamics. Cambridge University Press, Cambridge

Hammond KA, Diamond J (1997) Maximum sustained energy budgets in humans and animals. Nature 386:457–462

8 Adaptive Strategies of Frugivore Bats to Andean Cloud Forests

Hayes JP, Garland T Jr (1995) The evolution of endothermy: testing the aerobic capacity model. Evolution 49(5):836–847

HerbstLH (1983) Nutritional analyses of the wet season diet of Carollia perspicillata (Chiroptera: Phyllostomidae) in Parque Nacional Santa Rosa, Costa Rica. Thesis, University of Miami

Herbst LH (1985) The role of nitrogen from fruit pulp in the nutrition of a frugivorous bat, *Carollia perspicillata*. Biotropica 18:39–44

Herreid CF, Kessel B (1967) Thermal conductance in birds and mammals. Comparative Biochem Physiol 21:405–414

Herrera L, Gutierrez E, Hobson KA, Altube B, Díaz WC, Sánchez-Cordero V (2002) Sources of assimilated protein in five species of new world frugivorous bats. Oecologia (Berlin) 133:280–287

Hosken DJ, Withers PC (1997) Temperature regulation and metabolism of an Australian bat, *Chalinolobus gouldii* (Chiroptera: Vespertilionidae) when euthermic and torpor. J Comp Physiol B 167:71–80

Hosken DJ, Withers PC (1999) Metabolic physiology of euthermic and torpid lesser long-eared bats, *Nyctophilus geoffroyi* (Chiroptera: Vespertilionidae). J Mammal 80(1):42–52

IudicaCA (2000) Systematic revision of the neotropical fruit bats of the genus Sturnira: a molecular and morphological approach. Dissertation, University of Florida, FL,p 284

Kendeigh S (1970) Energy requirements for existence in relation to size of bird. Condor 72:60–65

Koopman KF (1978) Zoogeography of Peruvian bats with special emphasis of the role Andes. Am Mus Novit 2651:1–33

Koopman KF (1982) Biogeography of bats of South America. In: Mares MA, Genoways HH (eds) Mammalian biology in South America, vol 6. Pymatuning Laboratory of Ecology, University of Pittsburgh, Pittsburgh

Korine C, Zinder O, Arad Z (1999) Diurnal and seasonal changes in blood composition of the free-living Egyptian fruit bat (*Rousettus aegyptiacus*). J Comp Physiol B 169:280–286

Kovtun MF, Zhukova NF (1994) Feeding and digestive intensity of chiropterans of different trophic groups. Folia Zool 43:377–386

Law BS (1994) Climatic limitation of the southern distribution of the common blossom bat *Syconycteris australis* in New South Wales. Aust J Ecol 19:366–374

Linares OJ (1998) Mamíferos de Venezuela. Sociedad Conservacionista Audubon de Venezuela, Caracas

Lomolino MV (2001) Elevation gradients of species-density: historical and prospective views. Glob Ecol Biogeogr 10:3–13

Lovegrove BG (2005) Seasonal thermoregulatory responses in mammals. J Comp Physiol B 175:231–247

Mayr E (1963) Animal species and evolution. Harvard University Press, Cambridge

McManus JJ (1977) Thermoregulation. In: Baker RJ, JKJ J, Carter DC (eds) Biology of bats of the New World family Phyllostomidae, vol 2. Texas Tech Press, Lubbock

McNab BK (1969) The economics of temperature regulations in Neotropical bats. Comparative Biochem Physiol 31:227–268

McNab BK (1970) Body weight and the energetic of temperature regulation. J Exp Biol 53:329–348

McNab BK (1971) On the ecological significance of Bergmann's rule. Ecology 52(5):845–854

McNab BK (1973) Energetic and the distribution of vampires. J Mammal 54:131–144

McNab BK (1974) The behavior of temperate cave bats in a subtropical environment. Ecology 55:943–958

McNab BK (1976) Seasonal fat reserves o bats in two tropical environments. Ecology 57:332–338

McNab BK (1980a) Food habits, energetic, and the population biology of mammals. Am Nat 116(1):106–124

McNab BK (1980b) On estimating thermal conductance in endotherms. Physiol Zool 53(2):145–156

McNab BK (1982a) Evolutionary alternatives in the physiological ecology of bats. In: Kunz TH (ed) Ecology of bats. Plenum, New York

McNab BK (1982b) The physiological ecology of South American mammals. In: Mares MA, Genoways HH (eds) Mammalian biology in South America. Pymatuning Laboratory of Ecology, University of Pittsburgh, Pittsburgh

McNab BK (1983) Energetic, body size, and the limits to endothermy. J Zool 199:1–29

McNab BK (1986) The influence of food habits on the energetic of eutherian mammals. Ecol Monogr 56(1):1–9

McNab BK (1988) Complications inherent in scaling the basal rate of metabolism in mammals. Q Rev Biol 63:25–53

McNab BK (1989) Temperature regulation and rate metabolism in three bornean bats. J Mammal 70(1):153–161

McNab BK (2003) Standard energetic of phyllostomid bats: the inadequacies of phylogenetic-contrast analyses. Comparative Biochem Physiol 135:357–368

McNab BK, Armstrong MI (2001) Sexual dimorphism and the scaling of energetic in flying foxes of the genus *Pteropus*. J Mammal 82(3):709–720

McNab BK, Bonaccorso FJ (2001) The metabolism of new Guinean pteropodid bats. J Comp Physiol B 171:201–214

Molinari J, Soriano PJ (1987) Sturnira bidens. Mamm Species 276:1–4

Morrison DW (1980) Efficiency of food utilization by fruit bats. Oecologia (Berlin) 45:270–273

Owen JG (1990) Patterns of mammalian species richness in relation to temperature, productivity, and variance in elevation. J Mammal 71(1):1–13

Pacheco V, Patterson BD (1991) Phylogenetic relationships of the New World bat genus *Sturnira* (Chiroptera: Phyllostomidae). Bull Am Mus Nat Hist 206:101–121

Patterson BD, Pacheco V, Solari S (1996) Distributions of bats along an elevational gradient in the Andes of South-Eastern Peru. J Zool 240:637–658

Rahbek C (1995) The elevational gradient of species richness: a uniform pattern? Ecography 18:200–205

Rosenzweig ML (1992) Species diversity gradients: we know more and less than we thought. J Mammal 73:715–730

RuizA (2006) Termoregulación, recursos y límites altitudinales en murciélagos frugívoros y nectarívoros andinos. Dissertation, Postgrado en Ecología Tropical, Facultad de Ciencias, Universidad de Los Andes, Mérida, Venezuela, p 215

Sarmiento G (1986) Ecological features of climate in high tropical mountains. In: Vuilleuimier F, Monasterio M (eds) High tropical biogeography. Oxford University, New York

Schmidt-Nielsen K (1990) Physiology: adaptation and environment, 5th edn. Cambridge University Press, Cambridge

Schondube J, Martínez del Rio C (2004) Sugar and protein digestion in flower piercers and hummingbirds: a comparative test of adaptive convergence. J Comp Physiol B 174:263–273

Simmons NB (2005) Order Chiroptera. In: Wilson DE, Reeder DM (eds) Mammals species of the world: a taxonomic and geographic reference, 3rd edn. Johns Hopkins University Press, Baltimore

SorianoPJ (1983) La comunidad de quirópteros de las selvas nubladas en los Andes de Mérida. Patrón reproductivo de los frugívoros y las estrategias fenológicas de las plantas. Tesis de Magister Scientiae, Facultad de Ciencias, Universidad de Los Andes, Mérida

Soriano PJ (2000) Functional structure of bat communities in tropical rain forests and Andean cloud forests. Ecotropicos 13(1):1–20

Soriano PJ, Díaz de Pascual A, Ochoa J, Aguilera M (1999) Biogeographic analysis of the mammal communities in the Venezuelan Andes. Dermatol Int 24(1):17–25

Soriano PJ, Ruiz A, Arends A (2002) Physiological responses to ambient temperature manipulation by three species of bats from Andean cloud forest. J Mammal 83(2):445–457

Speakman JR, Thomas DW (2003) Physiological ecology and energetic of bats. In: Kunz TH, Fenton MB (eds) Bat ecology. University of Chicago Press, Chicago

Stevens RD (2004) Untangling latitudinal richness gradients at higher taxonomic levels: familial perspectives on the diversity of new world bat communities. J Biogeogr 31:665–674

8 Adaptive Strategies of Frugivore Bats to Andean Cloud Forests

Stevens RD (2006) Historical processes enhance patterns of diversity along latitudinal gradients. Proc Roy Soc 273:2283–2289

Studier EH (1970) Evaporative water loss in bats. Comparative Biochem Physiol 35:935–943

Studier EH, Wilson DE (1970) Thermoregulation in some neotropical bats. Comparative Biochem Physiol 34:251–262

Tamsitt JR, Cadena A, Villarraga E (1986) Records of bats (*Sturnira magna* and *Sturnira aratathomasi*) form Colombia. J Mammal 67(4):754–757

Terborgh J (1971) Distribution on environmental gradients: theory and a preliminary interpretation of distributional patterns in the avifauna of the cordillera Vilcabamba, Peru. Ecology 52(1):23–40

Terborgh J (1977) Bird species diversity on an Andean elevation gradient. Ecology 58:1007–1019

Thomas DW (1984) Fruit intake and energy budgets of frugivorous bats. Physiol Zool 57(4):457–467

Thomas DW, Cloutier D (1992) Evaporative water loss by hibernating little brown bats, *Myotis lucifugus*. Physiol Zool 65(2):443–456

Tilman D (1982) Resource competition and community structure. Princeton University Press, Princeton

Timm RM, LaVal RK (1998) A field key to the bats of Costa Rica. Occ Publ Ser 22:1–24

Tuttle MD (1970) Distribution and zoogeography of Peruvian bats, with comments on natural history. Univ Kansas Sci Bull 49:45–86

Villalobos F, Valerio AA (2002) The phylogenetic relationships of the bat genus *Sturnira* gray, 1842 (Chiroptera: Phyllostomidae). Mamm Biol 67:268–275

Willig MR, Bloch CP (2006) Latitudinal gradients of species richness: a test of the geographic area hypothesis at two ecological scales. Oikos 112:163–173

Willig MR, Patterson BD, Stevens RD (2003) Patterns of range size, richness, and body size in the Chiroptera. In: Kunz TH, Fenton MB (eds) Bat ecology. University of Chicago Press, Chicago

Zar JH (1999) Biostatistical analysis, 4th edn. Prentice Hall, New Jersey

Chapter 9
Neotropical Biodiversity: Hypotheses of Species Diversification and Dispersal in the Andean Mountain Forests

Angela M. Mendoza-Henao and Juan C. Garcia-R

9.1 Introduction

It is well established that species diversity is unevenly distributed across the globe (Wallace 1876) and a high proportion of it occurs in the tropics (Gaston 2000), particularly in the Andean mountain forests. The Andes hosts an exceptional diversity and endemism of vertebrate biota. More than 3400 vertebrate species (excluding fishes) are found in this mountain region and almost half of them are endemic (Myers et al. 2000; Orme et al. 2005; Elsen et al. 2018). The substantial number of species occurring in the Andes is a reflection of high beta diversity and spatial turnover at local and regional scales (Jankowski et al. 2009; Sklenář et al. 2014). However, this high biodiversity raises questions about the relevance of the different driving factors associated with species diversification and dispersal events.

Several hypotheses have been suggested to help explain the high diversity in the Andes (Table 9.1). Species richness and endemism are attributed to a mixture of historical, evolutionary, and ecological processes (Rahbek and Graves 2001; Rahbek et al. 2007, 2019a; Fjeldså et al. 2012; Rangel et al. 2018). The processes alone are not mutually exclusive but instead form a complex dimensional interrelationship for the observed spatial and temporal patterns of species diversity in the Andes region. Thus, environmental heterogeneity, in both time and space, and evolutionary history, seems to be relevant for understanding the origin and distribution of extant biodiversity (Rull 2011; Antonelli et al. 2018; Perrigo et al. 2020). The mechanisms of diversification differ based on the relative role of each process and the

---

A. M. Mendoza-Henao
Departamento de Zoología, Instituto de Biología, Universidad Nacional Autónoma de México, Mexico City, México

J. C. Garcia-R (✉)
Molecular Epidemiology and Public Health Laboratory, Hopkirk Research Institute, School of Veterinary Science, Massey University, Palmerston North, New Zealand
e-mail: j.c.garciaramirez@massey.ac.nz

© Springer Nature Switzerland AG 2021
R. W. Myster (ed.), *The Andean Cloud Forest*,
https://doi.org/10.1007/978-3-030-57344-7_9

Table 9.1 Some of the hypotheses that explain the biodiversity in the Andes region

Category	Hypothesis	Summary description	References
Evolutionary	Time-for-speciation (mountain museum hypothesis)	Earlier colonization of regions or habitats facilitates higher richness because they have more time for in situ speciation and accumulation of species	Stephens and Wiens (2003)
	Diversification rate (species pump)	Faster diversification rates (i.e., speciation minus extinction over time) of clades in some regions and habitats can generate higher species richness	Smith et al. (2007)
Spatial and ecological	Pleistocene oscillations	Pleistocene oscillations caused fragmentation of species ranges and local adaptation of populations	Turchetto-Zolet et al. (2013)
	Andean orogenesis	Andean uplift created isolated "sky islands" surrounded by drier valleys, which act as barriers to the dispersal of species	Hoorn et al. (2010)
	Across-Andes dispersal	Amazonian species episodically "spilled over" into western South America during humid periods, across low passes in the Andean region	Haffer (1974)
	Environmental heterogeneity	Heterogeneous environments at mid-elevations can harbor more species, enhance species persistence, and promote adaptive radiations because they can have a rich array of suitable conditions	Rosenzweig (1995), Fjeldsaå and Lovett (1997), Sedano and Burns (2010)
	Climatic zonation by thermal tolerances	Narrow thermal tolerances of tropical species and greater climatic stratification of tropical mountains create opportunities for climate-associated parapatric and allopatric speciation	Janzen (1967)
	Climatic-niche divergence	Accelerated rates of climatic-niche evolution in montane areas are associated with increased diversification rates	Kozak and Wiens (2010)

lineage-specific attributes. However, the precise association of processes and traits is often unclear.

Multiple competing hypotheses attempt to explain the diversity of vertebrates in the tropical montane forest of the Andes. Proposed drivers of diversification in the tropical Andes can be grouped into spatial (e.g., species–area relationship, mid-domain effect), climatic and topographic (e.g., temperature, topographical gradients), historical (e.g., speciation rates, clade age) and biotic (e.g., competition, thermal tolerance) hypotheses. In this chapter, we summarize some of the leading hypotheses explaining the origin and distribution of vertebrate clades in the Andean montane forests. Nonetheless, these are non-exclusive hypotheses to vertebrates and can also be used to explain the high diversity among other groups in the tropical Andes, including invertebrates and plants.

9.2 Drivers of Species Diversification in the Tropical Andes

9.2.1 Shape

The simplest explanation to the observed diversity in the montane forest of the Andes, without the need to call for any particular evolutionary nor geologic process, is the spatial pattern (Stein et al. 2014; Rahbek et al. 2019b). The spatial hypothesis suggests that a reduction in diversity with altitude is only a consequence of more extensive areas at low elevations than the available area at high elevations. The Species–Area Relationships (SAR) assumes that more area can bear more species (Rosenzweig 1995). For example, Rahbek (1997) found that species richness of birds in South America is controlled by area and remained nearly constant from lowlands up to 2600 m. However, more recent studies have shown that these lineal trends are obtained when the extent of the gradients sampled do not cover all the mountain system (Nogués-Bravo et al. 2008).

Several lineages of vertebrates in the mountain regions exhibit a hump-shaped or mid-elevational species richness peak (Kattan and Franco 2004; Meza-Joya and Torres 2016; Hutter et al. 2017; Quintero and Jetz 2018). The pattern has been observed in many mountain systems. Grinnell and Storer (1924) showed that several groups of terrestrial vertebrates have a unimodal richness pattern with the highest species richness about a third of the way up the Yosemite mountain. Nevertheless, this pattern is less likely to be found for small-ranged species (Colwell and Hurtt 1994; Colwell et al. 2004). An explanation for the observed richness pattern is the mid-domain effect (MDE) hypothesis. This hypothesis states that any random distribution of species on a bounded map (such as an elevational gradient) produces a peak of species richness near the center due to geometric boundary constraints (Colwell et al. 2004). The model predicts a parabolic curve of species richness, with a peak of 0.5 times the total number of species at the center of the gradient (Bokma et al. 2001). Despite several studies supporting the mid-domain effect, others have found that the model is not a general explanation for the diversity patterns observed (Hawkins and Diniz-Filho 2002; Dunn et al. 2007) but it is still used as a null hypothesis to test more complex scenarios.

9.2.2 Climate

Another important component influencing Andean biodiversity is the geographical location in the tropical region. The tropics are favored by gathering direct solar energy throughout the year, which provides a stable condition and higher annual productivity. However, the climate of tropical mountain regions is fundamentally different in complexity and variety than adjacent lowland regions. The uniquely heterogeneous environment in mountain systems plays a key role in generating and maintaining high diversity (Antonelli et al. 2018). The Northern Andes, for

a) Climatic heterogeneity

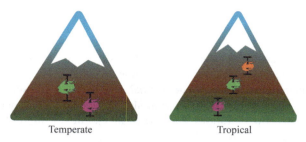

b) Topographic unevenness

Fig. 9.1 Processes of diversification in Andean biota caused by climate heterogeneity and topographic unevenness. (**a**) The elevational zonation of mountains in tropical regions presents more stable climatic conditions than temperate mountains, enabling small-ranged species and, therefore, more diversity; (**b**) The presence of multiple peaks, steep ridges, and inter-valleys act as barriers to dispersal for highland species, allowing independent speciation events throughout the ranges

example, captures roughly half of the world's climate types in a relatively small region, much more than those captured in the entire Amazon (Rahbek et al. 2019b).

Temperature is one of the most studied climatic variables related to biodiversity patterns because it reflects species' environmental tolerances. The Janzen's hypothesis (Janzen 1967) provides an explanation for the effect of yearly temperature variation in the biodiversity of mountain systems. Tropical mountains show less climatic variation during the year but a great variety of vegetation belts in each altitudinal band than mountains in template regions (Fig. 9.1a). This variation represents a significant physiological barrier to dispersal because species would experience extreme conditions beyond their adaptive or acclimation capabilities (McCain 2009). Therefore, species living in tropical mountains show narrower range sizes than temperate species at the same elevation, allowing a high species turnover (Cadena et al. 2011; Fjeldså et al. 2012).

9.2.3 Topography

Topographic complexity plays a major role in the origin and maintenance of biodiversity of montane areas. The Andes show multiple peaks, steep ridges, and inter-valleys that act as barriers to dispersal (Fig. 9.1b). The topographic unevenness enables habitat partitioning among species, which favor independent speciation events at similar range slopes (Hazzi et al. 2018). This might explain the high beta diversity and endemism observed for low dispersal organisms such as anurans (Lynch and Duellman 1997; García-R et al. 2012; De la Riva 2020) and small-range clades of hummingbirds (McGuire et al. 2014). The diversity–environment correlation invokes a "pure ecological" explanation involving population-level processes tied to dispersal and aggregation.

9.2.4 Geological and Environmental Shifts

The orogeny of the Andes has long been identified as a major driver of speciation, extinction, and biotic interchange (Hoorn et al. 2010; Kattan et al. 2016; Smith et al. 2014). The uplift of the Andes in the late Eocene around 40 million years ago (Ma) created new montane environmental conditions that opened up opportunities for diversification. These new conditions allow the isolation and restriction of gene flow among populations at both sides of the mountain range (Fig. 9.2a). The relation between Andes uplift and species formation has been studied in multiple groups of birds (e.g., parrots, Ribas et al. 2007; tanagers, Sedano and Burns 2010; woodpeckers, Weir and Price 2011; hummingbirds, McGuire et al. 2014) and frogs (e.g., poison frogs, Santos et al. 2009; glassfrogs, Castroviejo-Fisher et al. 2014; direct-developing frogs, Mendoza et al. 2015). For instance, Mendoza et al. (2015), using phylogenetic and ancestral range analyses, reconstructed the role of Andean orogeny in the diversification of direct-developing frogs (*Pristimantis*) across the Neotropics. The genus-level phylogeny and ancestral range reconstruction suggest that the uplift between north and central Andes generated a bridge for dispersal of species and high speciation in mid-elevational ranges (1000–3000 masl).

Besides the tectonic uplift of the Andes, the mountain system suffered multiple glacial cycles during the Plio-Pleistocene (Gregory-Wodzicki 2000; Graham 2009; Rahbek et al. 2019a). The elevational limits of the mountain forest belts moved upwards and downwards, providing isolated habitats and corridors to populations distributed across both lowland and pre-Andean landscapes (Fig. 9.2b). The radiation of montane species could have resulted from interchanges between phases of isolation (prompting local speciation) and connectivity (triggering diversification through dispersal and settlement in new areas) (Flantua and Hooghiemstra 2018). As a result, species rapidly diversified (species pump) within a variety of optimum montane habitats. Shifting climates caused range retractions and expansions of mountain species with consequences in biotic interactions and diversification.

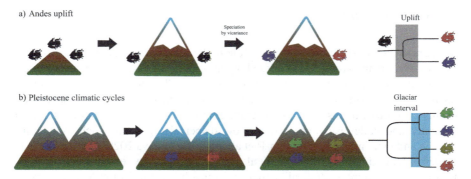

Fig. 9.2 Processes of diversification of Andean biota by orogenic and climatic changes in the Andes. (**a**) The Andean mountain uplift drives speciation (here illustrated by differences in the color of the frogs) by vicariance, fragmenting species distribution that was formerly continuous (**b**) Climatic dynamics during the Pleistocene facilitated temporal dispersal to new habitats and posterior isolation. Biological attributes and interactions (e.g., dispersal ability, thermal tolerance, and competition) are factors that determine the establishment in new habitats and successive genetic differentiation during climatic and geological events. Frogs, in this example, have features (e.g., dependence on water supply) that might limit species distribution and dispersal

(Ramírez-Barahona and Eguiarte 2013; Beckman and Witt 2015). Climatic cycles during the Pleistocene comprise one of the main sources of bird diversification in the Andes (Weir 2006; Gutiérrez-Pinto et al. 2012) and other mountain systems (Quintero and Jetz 2018).

9.2.5 Evolutionary History

Recent hypotheses have been tested to explain Andean diversity in terms of macroevolutionary processes. The time-for-speciation hypothesis suggests that intermediate elevations have higher richness because they were colonized earlier and had more time for speciation and species accumulation than those in other elevational zones (Stephens and Wiens 2003; Hutter et al. 2017). This scenario can occur when extinction rates are low, due to stable environmental conditions that favor an accumulation of highly adapted species. A second hypothesis, the species pump, states that higher species richness in mid-elevations is a by-product of faster diversification rates i.e., speciation minus extinction over time (Fjeldså 1994; Smith et al. 2007). This scenario is expected when selection for specialization across gradients (e.g., topographical heterogeneity) occurs, leading to a reduction in gene flow and subsequent speciation.

The time-for-speciation and species pump hypotheses are suggested as valid explanations for the distribution of diversity along elevational gradients in the Andes montane forest but direct evaluations are still required (Stephens and Wiens 2003; Hawkins et al. 2007, 2012; Smith et al. 2007). Based on a molecular divergence time estimation, Hutter et al. (2013) suggested that the pattern of mid-elevational diversity peak in Andean glassfrogs (Centrolenidae) is explained by the time-for-speciation hypothesis rather than diversification rates (Fig. 9.3). However, Hutter et al. (2017) found faster diversification rates in some Andean Hyloidea families when temporal and spatial scales were adjusted. Beckman and Witt (2015) showed that siskin (Aves: Fringillidae: *Spinus*) lineages in the Andes had higher diversification and dispersal rates than siskin lineages outside the Andes due to the expansion and contraction of the páramo during the Pleistocene climatic cycles. New approaches are now providing unprecedented means to quantify the dynamic processes of dispersal, speciation, and extinction rates, including phylogenetic information and comprehensive databases of geographic species records (e.g., López-Aguirre et al. 2018; Garcia-R et al. 2019).

9.2.6 Lineage Life-History Traits

Speciation events are linked to geophysical and climatic changes but the biological traits of the system under study need also be considered. The ability to disperse, colonize, and persist in a landscape matrix depends on biotic interactions (e.g., competition, predation) and physiological constraints (e.g., thermal tolerance) of the specific lineage (Hawkins et al. 2007, 2012; Graham et al. 2009; Pintanel et al. 2019). Different approaches have been used to understand the relative roles of life-history traits in dispersal, diversification and assembly of communities in the Andes region (Gillespie et al. 2012; Morlon 2014). The existence of greater thermal zonations in the Andes and niche conservatism in some tropical taxa limited the dispersal and facilitated allopatric isolation that, in turn, results in higher levels of speciation and accumulation of species (Cadena et al. 2011; Hutter et al. 2013). Niche evolution has also been inferred as a major contributor to ecological speciation in tropical mountains after finding major niche shifts between biomes in different Andean clades (Graham et al. 2004; Kozak and Wiens 2010; Pintanel et al. 2019).

9.3 Conclusion

Biodiversity in the Andes results from a long and complex history mediated by ecological, historical, and evolutionary processes. Mountain building, climate cycling, and montane environmental gradients played an important part in species richness

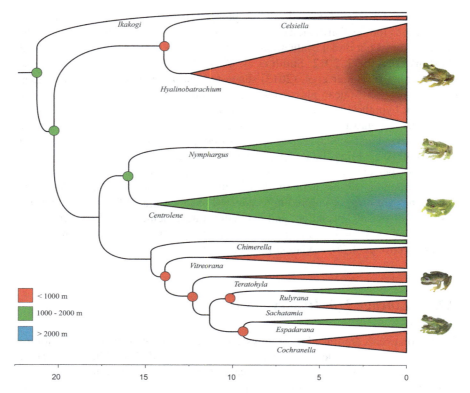

Fig. 9.3 Chronogram of glassfrogs (Anura: Centrolenidae). Recent studies suggest that glassfrogs were ancestrally present in mid-elevation habitats while lower and higher habitats were colonized more recently. Colored circles at each node represent the ancestral state reconstruction of elevational distributions. Ultrametric phylogenetic tree (time in Mega-annum or Ma) from Mendoza-Henao et al. (in prep) and altitudinal ancestral states from Hutter et al. (2013). Species at the right of the figure from up to down: *Hyalinobatrachium esmeralda*, *Nymphargus grandisonae*, *Centrolene hybrida*, *Teratohyla pulverata*, and *Espadarana prosoblepon*. Photos by AMMH

and spatial patterns in the Andes. Dispersal events are key promoters of speciation, while intrinsic factors (e.g., thermal tolerance) and biotic interactions determined differences in species diversity and distribution among clades. Further expansion and integration of different processes may offer a framework for understanding the patterns and evolution of diversity in the Andean mountain forests.

Acknowledgments AMMH is supported by the scholarship number CVU416922 from Consejo Nacional de Ciencia y Tecnología (CONACyT, Mexico), and Posgrado de Ciencias Biológicas of the Universidad Nacional Autónoma de México (UNAM). We thank Nicolas Hazzi for valuable comments.

References

Antonelli et al. (2018) Geological and climatic influences on mountain biodiversity. Nature Geoscience 11(10):718–725. https://www.nature.com/articles/s41561-018-0236-z#citeas

Beckman EJ, Witt CC (2015) phylogeny and biogeography of the new world siskins and goldfinches: rapid, recent diversification in the Central Andes. Mol Phylogenet Evol 87:28–45

Bokma F, Bokma J, Mönkkönen M (2001) Random processes and geographic species richness patterns: why so few species in the north? Ecography 24:43–49

Cadena CD, Kozak KH, Gómez JP, Parra JL, McCain CM, Bowie RCK, Carnaval AC, Moritz C, Rahbek C, Roberts TE, Sanders NJ, Schneider CJ, VanDerWal J, Zamudio KR, Graham CH (2011) Latitude, elevational climatic zonation and speciation in new world vertebrates. Proc R Soc B Biol Sci 279:720

Castroviejo-Fisher S, Guayasamin JM, Gonzalez-Voyer A, Vilà C (2014) Neotropical diversification seen through glassfrogs. J Biogeogr 41:66–80

Colwell RK, Hurtt GC (1994) Nonbiological gradients in species richness and a spurious rapoport effect. Am Nat 144:570–595

Colwell R, Xa K, Rahbek C, Gotelli N, Xa J (2004) The mid-domain effect and species richness patterns:what have we learned so far? Am Nat 163:1–23

De la Riva I (2020) Unexpected beta-diversity radiations in highland clades of andean terraranae frogs. In: Rull V, Carnaval AC (eds) Neotropical diversification: patterns and processes. Springer, Cham, pp 741–764

Dunn RR, McCain CM, Sanders NJ (2007) When does diversity fit null model predictions? Scale and range size mediate the mid-domain effect. Glob Ecol Biogeogr 16:305–312

Elsen PR, Monahan WB, Merenlender AM (2018) Global patterns of protection of elevational gradients in mountain ranges. Proc Natl Acad Sci 115:6004

Fjeldså J (1994) Geographical patterns for relict and young species of birds in Africa and South America and implications for conservation priorities. Biodivers Conserv 3:207–226

Fjeldså J, Bowie RCK, Rahbek C (2012) The role of mountain ranges in the diversification of birds. Annu Rev Ecol Evol Syst 43:249–265

Fjeldsaå J, Lovett JC (1997) Biodiversity and environmental stability. Biodivers Conserv 6:315–323

Flantua SGA, Hooghiemstra H (2018) Historical connectivity and mountain biodiversity. In: Hoorn AC (ed) Mountains, climate and biodiversity. Wiley-Blackwell, Oxford, pp 171–185

García-R JC, Crawford AJ, Mendoza ÁM, Ospina O, Cardenas H, Castro F (2012) Comparative phylogeography of direct-developing frogs (Anura: Craugastoridae: *Pristimantis*) in the Southern Andes of Colombia. PLoS ONE 7:e46077

Garcia-R JC, Gonzalez-Orozco CE, Trewick SA (2019) Contrasting patterns of diversification in a bird family (Aves: Gruiformes: Rallidae) are revealed by analysis of geospatial distribution of species and phylogenetic diversity. Ecography 42:500–510

Gaston KJ (2000) Global patterns in biodiversity. Nature 405:220–227

Gillespie RG, Baldwin BG, Waters JM, Fraser CI, Nikula R, Roderick GK (2012) Long-distance dispersal: a framework for hypothesis testing. Trends Ecol Evol 27:47–56

Graham A (2009) The Andes: a geological overview from a biological perspective. Ann Mo Bot Gard 96:371–385

Graham CH, Ron SR, Santos JC, Schneider CJ, Moritz C (2004) Integrating phylogenetics and environmental niche models to explore speciation mechanisms in dendrobatid frogs. Evolution 58:1781–1793

Graham CH, Parra JL, Rahbek C, McGuire JA (2009) Phylogenetic structure in tropical hummingbird communities. Proc Natl Acad Sci 106:19673

Gregory-Wodzicki KM (2000) Uplift history of the Central and Northern Andes: a review. Geol Soc Am Bull 112:1091–1105

Grinnell J, Storer TI (1924) Animal life in the yosemite: an account of the mammals, birds, reptiles, and amphibians in a cross-section of the Sierra Nevada. University of California Press, Berkeley

Gutiérrez-Pinto N, Cuervo AM, Miranda J, Pérez-Emán JL, Brumfield RT, Cadena CD (2012) Non-monophyly and deep genetic differentiation across low-elevation barriers in a Neotropical montane bird (Basileuterus tristriatus; Aves: Parulidae). Mol Phylogenet Evol 64:156–165

Haffer J (1974) Avian speciation in tropical South America with a systematic survey of the toucans (Ramphastidae) and the jacamars (Galbulidae). Nuttall Ornithological Club, Cambridge

Hawkins BA, Diniz-Filho JAF (2002) The mid-domain effect cannot explain the diversity gradient of Nearctic birds. Glob Ecol Biogeogr 11:419–426

Hawkins BA, Diniz-Filho JAF, Jaramillo CA, Soeller SA (2007) Climate, niche conservatism, and the global bird diversity gradient. Am Nat 170:S16–S27

Hawkins BA, McCain CM, Davies TJ, Buckley LB, Anacker BL, Cornell HV, Damschen EI, Grytnes J-A, Harrison S, Holt RD, Kraft NJB, Stephens PR (2012) Different evolutionary histories underlie congruent species richness gradients of birds and mammals. J Biogeogr 39:825–841

Hazzi NA, Moreno JS, Ortiz-Movliav C, Palacio RD (2018) Biogeographic regions and events of isolation and diversification of the endemic biota of the tropical Andes. Proc Natl Acad Sci 115:7985

Hoorn C, Wesselingh FP, ter Steege H, Bermudez MA, Mora A, Sevink J, Sanmartín I, Sanchez-Meseguer A, Anderson CL, Figueiredo JP, Jaramillo C, Riff D, Negri FR, Hooghiemstra H, Lundberg J, Stadler T, Särkinen T, Antonelli A (2010) Amazonia through time: andean uplift, climate change, landscape evolution, and biodiversity. Science 330:927–931

Hutter CR, Guayasamin JM, Wiens JJ (2013) Explaining Andean megadiversity: the evolutionary and ecological causes of glassfrog elevational richness patterns. Ecol Lett 16:1135–1144

Hutter CR, Lambert SM, Wiens JJ (2017) Rapid Diversification and time explain amphibian richness at different scales in the tropical andes, earth's most biodiverse hotspot. Am Nat 190(6):828–843

Jankowski JE, Ciecka AL, Meyer NY, Rabenold KN (2009) Beta diversity along environmental gradients: implications of habitat specialization in tropical montane landscapes. J Anim Ecol 78:315–327

Janzen DH (1967) Why mountain passes are higher in the Tropics. Am Nat 101:233

Kattan GH, Franco P (2004) Bird diversity along elevational gradients in the Andes of Colombia: area and mass effects. Glob Ecol Biogeogr 13:451–458

Kattan GH, Tello SA, Giraldo M, Cadena CD (2016) Neotropical bird evolution and 100 years of the enduring ideas of Frank M. Chapman. Biol J Linn Soc 117:407–413

Kozak KH, Wiens JJ (2010) Accelerated rates of climatic-niche evolution underlie rapid species diversification. Ecol Lett 13:1378–1389

López-Aguirre C, Hand SJ, Laffan SW, Archer M (2018) Phylogenetic diversity, types of endemism and the evolutionary history of New World bats. Ecography 41:1–12

Lynch JD, Duellman WE (1997) Frogs of the genus *Eleutherodactylus* in western Ecuador: systematics, ecology, and biogeography. Nat History Museum 23:1–236

McCain CM (2009) Global analysis of bird elevational diversity. Glob Ecol Biogeogr 18:346–360

McGuire JA, Witt CC, Remsen JV, Corl A, Rabosky DL, Altshuler DL (2014) Molecular phylogenetics and the diversification of hummingbirds. Curr Biol 24:9

Mendoza ÁM, Ospina OE, Cárdenas-Henao H, García-R JC (2015) A likelihood inference of historical biogeography in the world's most diverse terrestrial vertebrate genus: diversification of direct-developing frogs (Craugastoridae: *Pristimantis*) across the Neotropics. Mol Phylogenet Evol 85:50–58

Meza-Joya FL, Torres M (2016) Spatial diversity patterns of Pristimantis frogs in the tropical Andes. Ecol Evol 6(7):1901–1913

Morlon H (2014) Phylogenetic approaches for studying diversification. Ecol Lett 17:508–525

Myers N, Mittermeier RA, Mittermeier CG, da Fonseca GAB, Kent J (2000) Biodiversity hotspots for conservation priorities. Nature 403:853–858

Nogués-Bravo D, Araújo MB, Romdal T, Rahbek C (2008) Scale effects and human impact on the elevational species richness gradients. Nature 453:216–219

Orme CDL, Davies RG, Burgess M, Eigenbrod F, Pickup N, Olson VA, Webster AJ, Ding T-S, Rasmussen PC, Ridgely RS, Stattersfield AJ, Bennett PM, Blackburn TM, Gaston KJ, Owens IPF (2005) Global hotspots of species richness are not congruent with endemism or threat. Nature 436:1016–1019

Perrigo et al. (2020) Why mountains matter for biodiversity. J Biogeogr 47(2):315–325. https://onlinelibrary.wiley.com/doi/10.1111/jbi.13731

Pintanel P, Tejedo M, Ron SR, Llorente GA, Merino-Viteri A (2019) Elevational and microclimatic drivers of thermal tolerance in Andean Pristimantis frogs. J Biogeogr 46:1664–1675

Quintero I, Jetz W (2018) Global elevational diversity and diversification of birds. Nature 555:246

Rahbek C (1997) The relationship among area, elevation, and regional species richness in neotropical birds. Am Nat 149:875–902

Rahbek C, Graves GR (2001) Multiscale assessment of patterns of avian species richness. Proc Natl Acad Sci U S A 98:4534–4539

Rahbek C, Gotelli NJ, Colwell RK, Entsminger GL, Rangel TFLVB, Graves GR (2007) Predicting continental-scale patterns of bird species richness with spatially explicit models. Proc R Soc Lond Ser B Biol Sci 274:165–174

Rahbek C, Borregaard MK, Antonelli A, Colwell RK, Holt BG, Nogues-Bravo D, Rasmussen CMØ, Richardson K, Rosing MT, Whittaker RJ, Fjeldså J (2019a) Building mountain biodiversity: geological and evolutionary processes. Science 365:1114

Rahbek C, Borregaard MK, Colwell RK, Dalsgaard B, Holt BG, Morueta-Holme N, Nogues-Bravo D, Whittaker RJ, Fjeldså J (2019b) Humboldt's enigma: what causes global patterns of mountain biodiversity? Science 365:1108

Ramírez-Barahona S, Eguiarte LE (2013) The role of glacial cycles in promoting genetic diversity in the neotropics: the case of cloud forests during the last glacial maximum. Ecol Evol 3:725–738

Rangel TF, Edwards NR, Holden PB, Diniz-Filho JAF, Gosling WD, Coelho MTP, Cassemiro FAS, Rahbek C, Colwell RK (2018) Modeling the ecology and evolution of biodiversity: biogeographical cradles, museums, and graves. Science 361:5452

Ribas CC, Moyle RG, Miyaki CY, Cracraft J (2007) The assembly of montane biotas: linking Andean tectonics and climatic oscillations to independent regimes of diversification in Pionus parrots. Proc R Soc B Biol Sci 274:2399–2408

Rosenzweig ML (1995) Species diversity in space and time. Cambridge University Press, Cambridge

Rull V (2011) Neotropical biodiversity: timing and potential drivers. Trends Ecol Evol 26:508–513

Santos JC, Coloma LA, Summers K, Caldwell JP, Ree R, Cannatella DC (2009) Amazonian amphibian diversity is primarily derived from Late Miocene Andean lineages. PLoS Biol 7:e1000056

Sedano RE, Burns KJ (2010) Are the Northern Andes a species pump for Neotropical birds? Phylogenetics and biogeography of a clade of Neotropical tanagers (Aves: Thraupini). J Biogeogr 37:325–343

Sklenář P, Hedberg I, Cleef AM (2014) Island biogeography of tropical alpine floras. J Biogeogr 41:287–297

Smith SA, Montes de Oca AN, Reeder TW, Wiens JJ (2007) A phylogenetic perspective on elevational species richness patterns in Middle American treefrogs: why so few species in lowland tropical rainforests? Evolution 61:1188–1207

Smith BT, McCormack JE, Cuervo AM, Hickerson MJ, Aleixo A, Cadena CD, Perez-Eman J, Burney CW, Xie X, Harvey MG, Faircloth BC, Glenn TC, Derryberry EP, Prejean J, Fields S, Brumfield RT (2014) The drivers of tropical speciation. Nature 515:406–409

Stein A, Gerstner K, Kreft H (2014) Environmental heterogeneity as a universal driver of species richness across taxa, biomes and spatial scales. Ecol Lett 17:866–880

Stephens PR, Wiens JJ (2003) Explaining species richness from continents to communities: the time-for-speciation effect in emydid turtles. Am Nat 161:112–128

Turchetto-Zolet AC, Pinheiro F, Salgueiro F, Palma-Silva C (2013) Phylogeographical patterns shed light on evolutionary process in South America. Mol Ecol 22:1193–1213

Wallace A (1876) The geographical distribution of animals, with a study of the relations of living and extinct faunas as elucidating the past changes of the earth's surface. Macmillan and Co., London

Weir JT (2006) Divergent timing and patterns of species accumulation in lowland and highland Neotropical birds. Evolution 60:842–855

Weir JT, Price M (2011) Andean uplift promotes lowland speciation through vicariance and dispersal in Dendrocincla woodcreepers. Mol Ecol 20:4550–4563

Chapter 10
Mapping Hydrological Ecosystem Services and Impacts of Scenarios for Deforestation and Conservation of Lowland, Montane and Cloud-Affected Forests

Mark Mulligan

10.1 Introduction

Ecosystem services are the benefits that human populations derive from nature, but which are not usually accounted for in the economic system that we operate within (see Fisher et al. 2009). For many of these services the local proximity of beneficiaries is important to the benefits received, for example, hazard mitigation benefits (such as erosion control and flood mitigation) are received downhill of the systems that provide them; hydrological and sediment retention benefits are received nearby downstream and aesthetic and nature-based tourism tend to benefit local populations (through economic opportunities) and tourist-facing local businesses.

Cloud forests, though small in global surface area, comprising between 2.21M km^2 (Mulligan 2010a) and 215,000 km^2 (Bubb et al. 2004) depending on the definition used, are found in some important headwater areas with very significant populations locally and downstream. This context is thought to be very different from that of the much more extensive lowland forests, which are considered to have relatively few local and downstream populations. Cloud forests may thus provide important ecosystem services that are disproportionate to their relatively small global extent. This paper uses a range of spatial data to examine the key ecosystem services provided by the global cloud forest estate relative to other forest types, the benefits provided, who benefits and what is necessary to protect these benefits in the future, continent by continent. We also examine some important considerations which determine how accessible these services may be downstream.

Living forests are considered to provide a range of ecosystem services as well as harbouring much of the world's diversity and rarity of species. The key services provided include:

M. Mulligan (✉)
Department of Geography, King's College London, London, UK
e-mail: mark.mulligan@kcl.ac.uk

© Springer Nature Switzerland AG 2021
R. W. Myster (ed.), *The Andean Cloud Forest*,
https://doi.org/10.1007/978-3-030-57344-7_10

(a) carbon storage and sequestration which mitigate against climate change (Naidoo et al. 2008),
(b) the provision and regulation of supplies of clean water through nutrient, contaminant and sediment retention (Brauman et al. 2007),
(c) the mitigation of environmental hazards such as landsliding, sediment transport, flooding and low flows (De Groot et al. 2010), as well as
(d) the provision of livelihood opportunities through ecotourism and the availability of non-timber forest products such as firewood, fruits, nuts, fungi, building materials and medicinal products (Jenkins et al. 2004).

Some of these services are realised by beneficiaries throughout the world, for example, through climate change mitigation provided by carbon storage and sequestration. Others are more local in their beneficiaries, such as the downslope beneficiaries of hazard mitigation and downstream beneficiaries of water services. Still others have local first benefits (for example, improved water production sustaining agriculture) but then teleconnections through commodity supply chains to second beneficiaries of this productivity throughout the world (see Mulligan 2015a).

Here we use a range of global datasets to map the ecosystem service provision of all tropical (ALLF), tropical montane (TMF), cloud-affected tree cover (CAFs) and cloud-affected forests (CFs). The latter two may be broadly considered cloud forests similar to the definitions adopted by Mulligan (2010a) and Bubb et al. (2004), respectively. Our objective is to examine the significance of cloud-affected forest types in the provision of key ecosystem services, how well they are protected and the tropical forest conservation foci that would be necessary to protect the ecosystem services provided whilst also sustaining the maximum tropical forest biodiversity. We examine both cloud-affected tree cover and cloud-affected forests since the boundary of the cloud forest condition is unclear and it is important to understand the extent to which cloud affected systems of different tree cover fractions (and thus forest/non-forest definitions) impact on ecosystem service delivery.

10.2   Materials and Methods

Here we map the provision of key ecosystem services by all tropical (ALLF), tropical montane forests (TMF), tropical montane cloud-affected forests (CAFs) and (the more extensive/intact) tropical montane cloud forests (CFs). ALLF are defined as all areas in the continents of South America, Central America, Africa and tropical and subtropical Asia (to 38°N) with >40% tree cover according to (Townshend et al. 2011) and thus include (seasonally) dry forests as well as tropical humid forests. Montane forests (TMF) are forests on these same continents with elevation >500 masl (according to Farr and Kobrick 2000). This is a simple definition of montane compared with many (*cf.* Funnell and Price 2003) but is considered relevant here because of the strong climatic influence on cloud forests (Jarvis and Mulligan 2009). CAFs are cloud-affected forests according to Mulligan (2010a) remote-sensing based hydroclimatic analysis, that is, all areas with forest cover >10% and annual mean fog frequency >70%, so including very fragmented forest cover (but still forests according to the FAO definition of tree cover >10% FAO (1998)). Cloud forests (CFs) are also defined according to Mulligan (2010a) but as areas with forest cover >40% and annual mean fog frequency >70% so only more extensive and intact forest is included commensurate with other analyses of cloud forest such as Bubb et al. (2004).

The analysis of ecosystem services conducted here is performed using the public domain Co$ting Nature platform (Mulligan 2015b; Mulligan et al. 2010; Mulligan and Clifford 2015; http://www.policysupport.org/costingnature) and associated global datasets as described below. The analysis is carried out continent by continent for South America, Central America, Africa and tropical and subtropical Asia, according to the hydrological continents defined by Lehner et al. (2008).

10.2.1 Mapping Potential Beneficiaries

Potential beneficiaries of forest ecosystem services include local and downstream populations that may benefit from provisioning and local regulating services as well as downstream water infrastructure in the form of dams that could benefit from water quantity, quality, flow regulation as well as sediment retention services. Beneficiaries are calculated based on global datasets for the distribution of population (LandScan 2007), dams (Mulligan et al. 2011), accumulated downstream at 1 km resolution using the flow directions provided by Lehner et al. (2008). For each forest type we calculate the mean downstream population for all cells in that forest type. This reflects the mean number of people potentially affected hydrologically by the forested pixels, though we will see later that distance downstream is a critical factor in determining who actually benefits significantly. As a proxy for infrastructure that potentially benefits from these forests we count the average number of downstream dams for each forested area using the 36,000 dams mapped by Mulligan et al. (2011).

10.2.2 Mapping Ecosystem Service Provision

Co$ting Nature uses a range of global datasets and simple models to map the sites of production for some key ecosystem services including carbon (storage and sequestration), water (quantity and quality), hazard mitigation (for landslide, drought, flood, coastal inundation) and nature-based tourism. It also uses a variety of species range data to calculate richness for the sampled red list species of mammals, amphibians, reptiles and birds. In each case the realised and potential services are mapped. A potential service is defined here as a service produced but not (currently) consumed, whereas a realised service is produced and available to current beneficiaries (Mulligan and Clifford 2015). For example, all potential carbon services are also realised because all carbon storage and sequestration contribute to climate change mitigation, with beneficiaries worldwide. However, where forests provide potential hydrological services, these are only realised as services if there are human beneficiaries (people, dams, irrigation projects) nearby downstream of these forests who can benefit from the services provided. The further downstream the potential beneficiaries from a site of ecosystem service production, the less their services will be influenced by that site.

10.2.3 Mapping Realised Carbon Services

The Co$ting Nature carbon storage and sequestration service is calculated as the combination of carbon stocks and carbon sequestration. Carbon sequestration is calculated here from the dry matter productivity (DMP) analysis of Mulligan (2009a) in which SPOT-VGT DMP calculated every 10 days at 1 km resolution on the basis of change in NDVI was averaged over the period 1998–2008, globally. DMP (t biomass/ha/yr) is multiplied by 0.42 (Ho 1976) to convert to units of tC/ha/yr. Above-ground carbon stock is calculated from Saatchi et al. (2011) for the areas in which data are available and Ruesch and Gibbs (2008) elsewhere. This is combined with soil carbon calculated from the map of Scharlemann et al. (2009) to produce a total above- and below-ground carbon stocks.

10.2.4 Mapping Realised Water Services

Water is considered a provisioning service in Co$ting Nature, though it also plays a role in the regulating services (see the section on hazard mitigation). Potential water services are measured as the volume of runoff (rainfall minus evapotranspiration) whose quality is unaffected by human activity, cumulated downstream. This is an indicator of the volume of clean water produced by a pixel. The human footprint on water quality index (HF) (Mulligan 2009b) is used as the indicator of water quality. The HF considers particular land uses to have the potential to contaminate water with sediment, agrochemicals, manures, etc. Land uses such as unprotected agriculture and pasture are considered non-point sources and roads, mines, oil and gas wells and urban areas are considered point sources. Agriculture in protected areas is considered to have a human footprint of zero as are areas with no human land use. The HF index multiplies the water balance of a pixel by the fractional cover of point and non-point sources in that pixel and cumulates this "polluted" water downstream using a streamflow network. The total volume of water flowing is also cumulated downstream and the HF in a pixel becomes the volume of polluted water as a percentage of the total water flowing. The nonpolluted volume of water expressed as a fraction is considered the potential water service of a pixel. Realised water services are calculated considering the use of these potential water services which, in turn, depends on the distribution of beneficiaries for hydrological services.

Maps of globally normalised downstream population, irrigation area and number of dams are pre-calculated to represent the distribution of beneficiaries. The population is derived from LandScan (2007) and is summed downstream at 1 km resolution using a drainage network derived from HydroSHEDS (Lehner et al. 2008) with the downstream total for a pixel assigned to that pixel. For number of dams we use the global dam census of Mulligan et al. (2011) and for irrigated land we cumulate the downstream irrigated areas of Siebert et al. (2007) for each point. These provide pixel-level indicators of the distribution of the beneficiaries of hydrological ecosystem services at a much finer resolution than previous global studies. Potential water services are multiplied by the normalised sum of all downstream beneficiaries to calculate the realised water services index.

10.2.5 Mapping Realised Hazard Mitigation Services

Hazard mitigation services are perhaps the most complex to assess since they are a function of:

(a) the potential for multiple hazards to occur and the human and infrastructural exposure to those hazards and vulnerability to the negative impacts of hazards. Exposure and vulnerability together define the risk.
(b) the role of local, upstream or near-coast ecosystems in reducing the potential impact of hazards (i.e. potential hazard mitigation services).

Hazard mitigation ecosystem services are then realised in those areas in which ecosystems provide hazard mitigation services but where there is also a risk. Areas with no risk may receive potential hazard mitigation services but these services are not realised. Co$ting Nature considers hazard potential as:

(a) the normalised frequency of cyclones according to Dilley (2005) multiplied by the normalised water balance as an index of high magnitude rainfall event hazards;
(b) for coastal inundation hazards we calculate distance from coast according to USGS (2006) and consider all pixels within 2 km as coastal. We also produce an index of low-lying land as all areas from 0 to 30 m according to the SRTM digital elevation model post-processed by Lehner et al. (2008). The probability of (coastal) inundation hazard is considered proportional to the normalised probability of Tsunami (according to NGDC 2011), cyclones and climatic sea level rise (considered for simplicity as equally likely everywhere) for all coastal areas. The probability of inundation hazard is the combination of these effects.
(c) for landslide hazards we consider the probability of landslides to increase with the normalised mean upstream slope gradient. Upstream slope gradient is

pre-calculated using the 1 km resolution digital elevation model and flow network of Lehner et al. (2008).

(d) the potential for flooding is considered proportional to normalised water balance with small potential in dry areas and high potential where water is plentiful. Though many floods are fluvial in nature, we use water balance rather than runoff in recognition that floods also occur from overwhelmed urban drainage, groundwater flooding and rainfall intensity greater than infiltration rates. These latter types of flood can be somewhat more predictable than fluvial floods.

Hazard potential is thus the mean of cyclone, inundation, landslide and flood probabilities, normalised either locally or globally as defined by the user. In addition to hazard potential, we consider hazard exposure as the exposure of human populations, activity and infrastructure. Socio-economic exposure is calculated as normalised GDP for 1990 (CIESIN 2002), population (LandScan 2007), agriculture (cropland and pasture fractional areas from Ramankutty et al. 2008) and infrastructure. Infrastructural exposure is calculated as the sum of the presence of dams (Mulligan et al. 2011), mines (Mulligan 2010b), oil and gas (Mulligan 2010c), urban areas (Schneider et al. 2009) and roads (FAO 2010). Exposure is multiplied by hazard potential to produce the index of exposure to hazards. Vulnerability to hazards is considered to scale with normalised GDP and infrastructure: the greater the GDP and infrastructure, the greater the capacity to cope with hazards. Risk is then exposure multiplied by the vulnerability.

Potential hazard mitigation *services* are then calculated according to a series of assumptions, based on knowledge of how ecosystems mitigate these hazards. We assume that landslide impacts at a point are mitigated according to the proportion of upstream area that is tree covered (using tree cover data from Hansen et al. 2006) or is protected (according to WDPA 2014). This is because tree cover reduces potential soil waterlogging and has been shown to reduce landslide frequency (Dapples et al. 2002) and protected areas will tend to have a lower agricultural and infrastructural impact—both of which can lead to increased frequency of landslides. Regulation of drought hazards (for example, reduced dry season flows) is assumed proportional to tree cover upstream. Although trees evaporate significant volumes of water and thereby reduce flows, they also encourage infiltration that helps to maintain dry season flows (Pena-Arancibia et al. 2012). Flood hazards are mitigated according to the proportion of upstream area that provides flood storage in the form of trees, wetlands (Lehner and Doll 2004), water bodies (USGS 2006) and floodplains (Mulligan 2010b). Mitigation from coastal inundation is considered to be provided by wetlands and mangroves (Spalding et al. 1997) but only in low-lying and coastal areas. Where these ecosystems occur inland they are assumed to have no coastal inundation mitigation potential. The total potential hazard mitigation services are then the mean of coastal flood regulation and landslide mitigation services.

Realised hazard mitigation services are calculated as the minimum of risk and potential hazard mitigation services for areas where risk is greater than 0. In other words if hazard mitigation potential is greater than risk, then hazard mitigation potential equals risk and the remaining hazard mitigation potential is unused. If risk is greater than hazard mitigation potential, then realised hazard mitigation is equal to potential hazard mitigation and some risk remains unmitigated.

10.2.6 Mapping Species Richness

We use the IUCN sampled red list extent of occupancy (EOO) data for amphibians (IUCN et al. 2008), mammals (IUCN et al. 2008), reptiles (IUCN 2010) and birds (Birdlife 2012) and calculate a measure of richness (total number of sampled species within each 10 km cell) combining all sampled species. We would have liked to incorporate similar data for plants but these data are not in the public domain. We calculate mean richness per forest type and continent as a simple arithmetic mean for all relevant pixels.

Having mapped these services we then subset them for continental zones of interest (ZOIs) representative of the four forest types: LRF, TMF, CAF and CF as defined above and present statistics for the total (sum) and mean (density) of each service for each forest cover type by continent. We then examine the distribution of protected areas within these forest types, by continent, using the WDPA August 2014 update (WDPA 2014) and finally the proportion of the realised ecosystem services provided by each forest type that are derived from land under protected status.

10.2.7 Distance Decay of Hydrological Services

Finally we examine the distance decay of hydrological ecosystem service from cloud forests using a new technique termed the hydrological footprint. Based on the human footprint on water quality methodology (Mulligan 2009b), the hydrological footprint (HF) is a measure of the influence of an area on flows downstream of that area. The HF is calculated using Co$ting Nature as the percentage of water in each pixel derived from a particular area of interest upstream (in this case ALLF, MF, CAF, CF or protected CAF). The calculation is performed by accumulating downstream the water balance derived from the area of influence (WBi) and accumulating the water balance of all areas (WBa). The HF in a given pixel is then (WBi/WBa)*100. The accumulation is done using a D8 accumulation function (see Jones 2002) and the water balance calculated monthly and annually from an interpolated rainfall dataset (Hijmans et al. 2005) minus a remotely sensed actual evapotranspiration climatology based on Mu et al. (2011). The hydrological footprint is thus a water-balance weighted measure of the distance decay of hydrological influence of an upstream area. The downstream influence thus depends on the size, distance upstream and water balance of the areas of interest *in relation to* surrounding areas between the upstream area of interest and the downstream site of interest.

The hydrological footprint is a simple indicator but does not take into account the fact that cloud forests provide additional inputs of water in the form of fog interception that are not usually available outside of the CAF zone. CAFs will also tend to have lower evapotranspiration than other forest types because of the frequent cloud cover (Bruijnzeel et al. 2011). The hydrological footprint may thus underestimate the influence of CAF. We test this for Costa Rica and Madagascar by running the WaterWorld model (Mulligan 2013) to calculate a water balance including fog contributions and then calculate the hydrological footprint using these data instead of the remote sensing data. This is done only for Costa Rica since computing limitations preclude it at the continental scale. We also utilise the WaterWorld "fog contribution to runoff" variable in this analysis as another measure of the decay of downstream influence, this time for the additional CAF input of fog only. Finally we calculate the rate of change in downstream decay with distance in order to better understand the distances within which downstream potential beneficiaries will be significantly influenced by the upstream CAF.

10.3 Results

10.3.1 The Distribution and Coverage of (Cloud-Affected) Forests

Global maps of the distribution of these four forest types are shown in Figs. 10.1, 10.2, 10.3 and 10.4 and statistics of their coverage by hydrological continent are given in Table 10.1. Clearly cloud forests (CFs) are generally the most restricted forest type, with lowland forests being the most extensive and montane and CAFs being in between. All continents have the presence of CAF and CF but they are particularly restricted in Africa and particularly extensive in Asia (Figs. 10.1, 10.2, 10.3 and 10.4).

Table 10.1 shows that forests constitute some 26 million square km in these continents, representing 26.9% of land. Of these forests, some 33% can be considered montane using the definition adopted here, 31% are considered cloud affected and 18% cloud forests. Proportions vary between continents with South America having the greatest percentage of territory under ALLF but tropical and subtropical Asia having the greatest percentage of territory under montane and cloud forest types at 10–15% compared with less than 10% for South and Central America. Cloud affected and cloud forests are particularly rare in Africa at 4.1% and 1.7% of land, respectively, though the corresponding land areas covered are significant given the size of the continent.

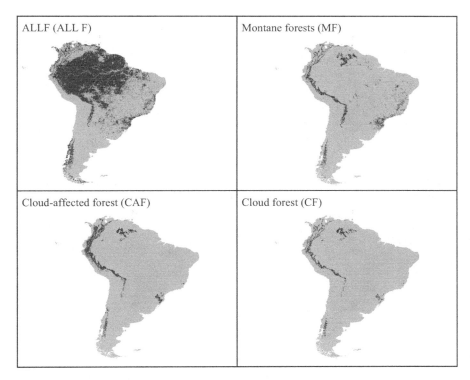

Fig. 10.1 The distribution of forest types used for ecosystem service comparison: South America

10.3.2 Potential Beneficiaries of (Cloud-Affected) Forest Ecosystem Services

Table 10.2 describes some key indicators of the density of potential beneficiaries by forest type and continent. These are presented as densities to indicate the mean number of people locally and downstream that are supported by the different forest types. For the benefitting population within the forested area, numbers are also multiplied by the area of each forest type from Table 10.1 to arrive at total number of beneficiaries. These are considered potential beneficiaries because, although ecosystem services may be provided, the extent to which they are realised will also depend upon economic, social and cultural influences on access and upon distance decay functions downstream.

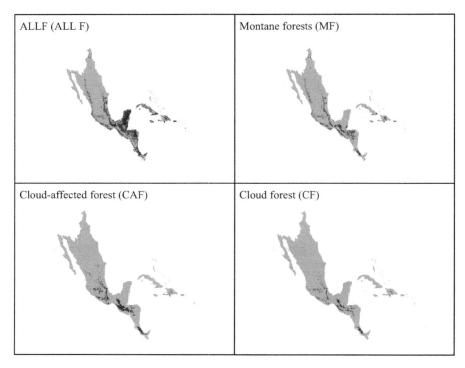

Fig. 10.2 The distribution of forest types used for ecosystem service comparison: Central America

10.3.3 Local Populations

For all continents population densities in forested areas tend to be lower than the continental average, though in African and Central American CAF areas, populations are higher than the continental average (perhaps because of the extensive low-population deserts in these regions). ALLF have a human population density between 2 and 32 persons per sq. km combining some 247M of the 6756M in the study region (3.6%). Population in montane forests are higher in Central America and South America but lower in Africa and Asia, combining 84.9M (1.3% of the total). Population densities in CAFs are much higher than ALLF or montane forests and sum to 246.4M (3.6% of the total) in only a third of the forest area occupied by ALLF. Indeed for Central America and Africa human population densities in CAFs are greater than the continental average. Population densities in CFs are more similar to those across ALLF but still represent 60.5M (0.8% of the study area population). Thus the local populations supported by these various forest types, especially CAFs and CFs are a very small proportion of the pantropical total populations.

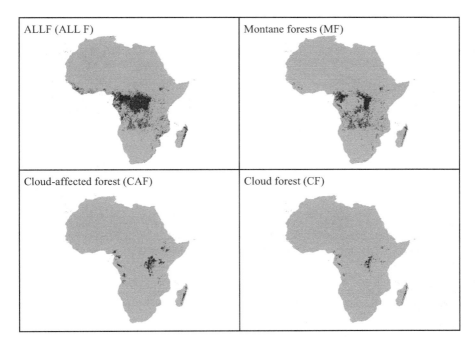

Fig. 10.3 The distribution of forest types used for ecosystem service comparison: Africa

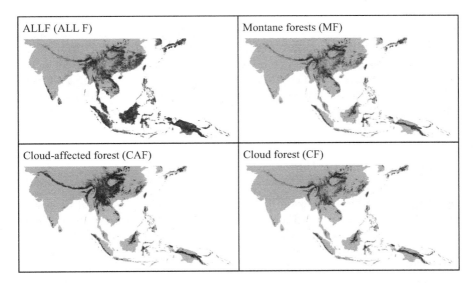

Fig. 10.4 The distribution of forest types used for ecosystem service comparison: tropical and subtropical Asia

Table 10.1 Coverage of forest types by continent

	ALLF	MF	CAF	CF
Definition	Tree cover >40%	Tree cover >40% and elevation >500	Mulligan (2010a) Tree cover >10% and fog frequency >70%	Mulligan (2010a) Tree cover >40% and fog frequency >70%
South America (SA)	7,037,844 km² (50.2%)	987,069 km² (7.4%)	909,329 km² (6.5%)	543,396 km² (3.9%)
Central America (CA)	590,472 (19.3%)	231208 (7.5%)	171,607 (5.4%)	92,971 (2.9%)
Africa (AF)	2,771,542 km² (19.8%)	1,581,671 km² (11.1%)	585,002 km² (4.1%)	253,017 km² (1.7%)
Asia	*6,232,291 km² (62.5%)*	*2,731,052 km² (28.6%)*	*3,327,390 km² (34.8%)*	*1,843,343 km² (17.2%)*
Asia (tropical)	3,655,757 km² (43.65%)	1,259,624 km² (15.04%)	1,169,172 km² (13.96%)	874,366 km² (10.44%)
Asia (tropics and subtropics to 38°) (ASTS)	5,762,456 (34.35%)	2,630,744 (15.68%)	3,434,530 (20.47%)	2,088,445 (12.44%)
Total (SA, CA, AF, ASTS)	16,162,314 (26.9% of land)	9,421,368 (33.6% of ALLF)	9,597,030 (31.6% of ALLF)	5,695,538 (18.4% of ALLF)

10.3.4 Downstream Populations

Numbers of downstream populations are an order of magnitude higher for forests in Africa and Asia compared with Central and South America. In all cases the mean number of persons downstream of forests increases from ALLF, through montane forests and peaks for CAFs. In Central America, Africa and Asia these have higher downstream populations than the all-land average. Forests have, on average, tens of thousands of downstream persons in Central and South America and hundreds of thousands in Africa and Asia.

10.3.5 Downstream Dams

Dams can be considered points in the landscape at which hydrological ecosystem services are monetized through the provision of irrigation, domestic and industrial water or hydropower. Number of downstream dams is highest in Asia and then Africa, followed by Central America and South America. Once again the mean number of dams downstream of forested pixels increases as we go from ALLF to MF with the peak number of dams downstream of CAFs. Each pixel of CAF serves an average of <0.4 dams in Central and South America, 0.4 dams in Africa and 1.0 dam in Asia. Whilst some pixels may have no downstream dams and other pixels may have many, the average is indicative of the overall influence of these forests on dams (Table 10.2).

10 Mapping Hydrological Ecosystem Services and Impacts of Scenarios...

Table 10.2 Beneficiaries in and downstream of forest types. Values greater than the all-land value for the given variable and continent are emboldened

Continent	Variable	ALLF	MF	CAF	CF	ALL
Central America	Mean population within forested area (person/km²)	19.9 (11.75M)	21.7 (5.01M)	**63.2** (10.85M) PROT: 30.6 (5.25M)	33.4 (3.1M)	48.0 (146M)
	Mean downstream population of forested area (person/km²)	13,890	26,473	32,866 PROT: **36,381**	26,582	36,079
	Mean downstream # dams of forested area	0.116	0.264	0.325 PROT: **0.427**	0.230	0.379
South America	Mean population within forested area (person/km²)	2.4 (16.89M)	5.7 (5.62M)	17.5 (15.91M) PROT: 4.9 (4.4M)	5.5 (2.98M)	21.8 (305M)
	Mean downstream population of forested area (person/km²)	18,518	34,578	33,311 PROT: 20,224	26,244	43,803
	Mean downstream # dams of forested area	0.022	0.099	0.060 PROT: **0.126**	0.059	0.108
Africa	Mean population within forested area (person/km²)	11.6 (32.1 M)	10.0 (15.8 M)	**86.0** (50.3M) PROT: 36.0 (21 M)	24.0 (6.1M)	51.2 (716.7M)
	Mean downstream population of forested area (person/km²)	144,566	191,456	**882,667** PROT: **943,853**	404,480	861,804
	Mean downstream # dams of forested area	0.053	0.088	**0.418** PROT: **0.446**	0.165	0.284
Tropical and subtropical Asia	Mean population within forested area (person/km²)	32.4 (186.7M)	22.2 (58.4M)	49.3 (169.3M) PROT: 11.3 (38.8M)	23.1 (48.2M)	192.0 (5587M)
	Mean downstream population of forested area (person/km²)	323,289	539,555	**735,186** PROT: **584,742**	**609,744**	564,889
	Mean downstream # dams of forested area	0.343	0.603	**1.00** PROT: **0.86**	**0.69**	0.668
Combined	Mean population within forested area (person/km²)	247M (3.6% of study area total)	84.9M (1.3%)	246.4M (3.6%) PROT: (69.5M) (1.02%)	60.5M (0.8%)	6756M

10.3.6 Beneficiaries of Protected Cloud-Affected Forests

For CAFs we have also calculated the mean for each variable that occurs within protected CAFs. Population densities are lower in protected CAFs as might be expected (and which may reflect the limitations of the population data, in which protected areas are used as one of the variables in the disaggregation of administrative region level population data). Some 69.5M (1.02%) of the study area population are within protected CAFs. Downstream populations for *protected* CAFs vary from tens of thousands in Central and South American protected CAFs to hundreds of thousands for Asian and African *protected* CAFs. These are a small fraction of the millions to billions of people in these continents, all others are vulnerable to changing hydrological ecosystem services resulting from poor land management. Downstream dam densities for *protected* CAFs tend to be higher than for all CAFs, except for Asia, suggesting strategic protection of dam watersheds.

10.3.7 Mapping Ecosystem Service Provision by Different Forest Types

For each continent we now examine a range of ecosystem services (both potential and realised, *sensu* Mulligan 2015b) and provide mean provision and total provision within each forest cover type, alongside information on tree cover, biodiversity, protected areas and deforestation for context. This is used to examine the natural capital and ecosystem service provision status of different forest types in order to place cloud forests within context.

10.3.8 Central America

For Central America (Table 10.3), the mean per-unit-area clean water provision tends to be highest for CAF and CFs, since these are extensive on the continent. For realised water provision all types of forest have greater provision per-unit-area than the continent-wide average (ALL-L) and that is also true for all carbon services. Hazard mitigation services are greatest for MFs, CAFs and especially CFs. Species richness is also greatest for MF, CAF and especially CF. Protected area coverage is significant for all forest types, though the most extensive are for ALLF (72% of all protected areas). Deforestation rates are greater for ALLF than for montane forests and the lowest rates are in CFs. The rate in protected CAFs is only a little less than the all CAF rate. The 5.4% of the study area that are CAFs provide 13.8% of realised water, 8.9% of aboveground carbon, 7.7% of soil carbon, 12% of hazard mitigation and 7.9% of the total tree cover. Their absolute contribution is low relative to the more extensive ALLF class but for most services their provision per-unit-area is greater. Some 0.04 of 0.17 Mkm^2 of CAF (23.3%) are protected but this only leads to a reduction in deforestation rates of around 25% relative to unprotected CAFs. Protected CAFs secure less than half of

10 Mapping Hydrological Ecosystem Services and Impacts of Scenarios…

Table 10.3 Ecosystem service provision by different forest types, Central America

Ecosystem services sum (mean) (percent of all-land total)	ALLF	MF	CAF	CF	ALL-L
Water (potential Mm^3 clean water produced)	0.69 **(0.92)** [48.03%]	0.26 **(0.89)** [17.94%]	0.19 **(0.93)** [13.49%] PROT: [22.7%]	0.13 **(1.13)** [8.82%]	1.45 (0.37)
Water (realised Mm^3 clean water produced)	0.68 **(0.91)** [48.1%]	0.26 **(0.89)** [18.31%]	0.19 **(0.93)** [13.77%] PROT: [22.5%]	0.13 **(1.14)** [9.00%]	1.42 (0.36)
AG carbon (M tonnes C storage) (tonnes C/km^2)	7011 **(9308)** [39.8%]	2730 **(9381)** [15.5%]	1580 **(7519)** [8.98%] PROT: [20.7%]	1132 **(10052)** [6.43%]	17600 (4515)
Soil carbon (M tonnes C storage) (tonnes C/km^2)	8250 **(10952)** [23.2%]	3156 **(10845)** [8.9%]	2748 **(13074)** [7.7%] PROT: [14.4%]	1505 **(13359)** [4.2%]	35548 (9119)
Carbon sequestration (M Dg/ha/yr) (Dg/ha/yr)	12.14 **(16.12)** [34.2%]	4.9 **(16.7)** [13.7%]	3.2 **(15.35)** [9.1%] PROT: [15.2%]	1.9 **(16.6)** [5.3%]	35.4 (9.0)
Hazard mitigation index (0-1)	0.003 **(0.004)** [34.1%]	0.001 **(0.005)** [15.9%]	0.001 **(0.005)** [12.06%] PROT: [14.1%]	0.0006 **(0.006)** [7.09%]	0.009 (0.002)
Species richness (species)	**(418)** [22.7%]	**(425)** [8.9%]	**(422)** [6.4%] PROT: [13.68%]	**(436)** [3.5%]	(357)
Tree cover (M km^2) (%)	584 **(70)** [86.8%]	81.5 **(66)** [12.1%]	53 **(49)** [7.9%] PROT: [20.6%]	44 **(67)** [6.6%]	673 (40)
Protected area (M km^2) (%)	3.6 [72.4%]	0.45 [9.1%]	0.04 [6.5%]	0.25 [5%]	4.97
Deforestation (2000–2012) %	21.9 (2.59) [33.4%]	3.56 (2.86) [5.4%]	1.94 (1.78) [2.97%] PROT: (1.44) [20.8%]	1.25 (1.89) [1.91%]	65.5 (3.9)

CAF ecosystem services (Table 10.3), making the remaining services vulnerable in the high deforestation context of this region (Table 10.4).

10.3.9 South America

For South America (Table 10.4) the mean per-unit-area clean water provision tends to be highest for ALLF (because of the lack of polluting human influence on the lowland forests). For realised water provision all forest types have greater provision per-unit-area than ALL-L and that is also true for the carbon services (except sequestration in CAFs and CFs which is similar to, or lower than, the mean for ALL-L). Hazard mitigation services are greatest for CAFs and CFs. Species richness, on the other hand, is greatest for ALLF since it is dominated by high richness in extensive lowland forests on this continent. Protected area coverage is significant for all forest types. ALLF represent 72% of protected areas, whereas CAFs only 6.48%. Deforestation rates are greater in montane forests than lowlands but lowest in CAFs. The rate in protected CAFs is half that of the all CAF rate. The 6.5% of the study areas that are CAFs provide 6.94% of realised water, 6.94% of above-ground carbon, more than 8% of soil carbon, 11.45% of hazard mitigation and 7.92% of the total tree cover. Their absolute contribution is low relative to the much more extensive ALLF class and for most services their provision per-unit-area is less than this class, though CAFs and CFs provide higher per-unit-area hazard mitigation services than ALLF in South America. 0.32 of 0.91 Mkm^2 of CAF (35.2%) are protected leading to a halving of deforestation rates relative to unprotected CAFs and securing significant proportions of CAF biodiversity, tree cover and water/carbon ecosystem services in particular (Table 10.4).

10.3.10 Africa

For Africa (Table 10.5) the mean per-unit-area clean water provision tends to be highest for ALLF and CF (because of the relative lack of human influence on these forest types). For realised water provision all forest types have greater provision per-unit-area than ALL-L and that is also true for all carbon services (except soil carbon in ALLF and MFs, which is similar to or lower than the mean for ALL-L). Provided hazard mitigation services are greatest for CAFs. Species richness is greatest for CAFs, CFs and MFs, reflecting the geographical variety of these systems on the continent. ALLF represents only 11% of Africa's protected areas and CAFs only 2.2%, with many protected areas in Africa covering non-forest environments. Deforestation rates are greatest in ALLF and the rate in protected CAFs is a third that of the all CAF rate. The 4.1% of the study areas that are CAFs provide 6.4% of realised water, 6.2% of above-ground carbon, but only 5% of soil carbon, 4.3% of hazard mitigation but 7.6% of the total tree cover. Their absolute contribution is low relative to the much more extensive ALLF class and for most services their provision per-unit-area is less, though CAFs provide higher per-unit-area

10 Mapping Hydrological Ecosystem Services and Impacts of Scenarios...

Table 10.4 Ecosystem service provision by different forest types, South America

Ecosystem services sum (mean) [percent of all-land total]	ALLF sum (mean)	MF sum (mean)	CAF sum (mean)	CF sum (mean)	ALL-L sum (mean)
Water (potential Mm^3 clean water produced)[a]	10,956,727 (1.301) [64.23%]	1,204,232 (0.968) [7.15%]	1,160,024 (1.064) [6.89%] PROT: [35.6%]	802,928 (1.226) [4.78%]	16,801,766 (1.677)
Water (realised Mm^3 clean water produced)	10.778 **(1.280)**	1.181 **(0.950)**	1.149 **(1.054)** PROT: [35.0%]	0.794 **(1.214)**	16.477 (0.982)
AG carbon (M tonnes C storage) (tonnes C/km^2)[b]	114,352 **(13,586)** [64.32%]	14,281 **(11,482)** [7.13%]	10,966 **(10,059)** [6.94%] PROT: [43.6]	8,469 **(12,937)** [4.80%]	136,048 (8,109)
Soil carbon (M tonnes C storage) (tonnes C/km^2)	91,434 **(10,863)** [84.05%]	15,812 **(12,713)** [10.50%]	16,564 **(15,194)** [8.06%] PROT: [23.9%]	9,342 **(14,272)** [6.23%]	199,518 (11,892)
Carbon sequestration (M Dg/ha/yr) (Dg/ha/yr)	79,818 **(9,483)** [46.56%]	11,982 **(9,634)** [7.03%]	8,896 (8,160) [5.22%] PROT: [24.7%]	5,923 (9,048) [3.49%]	152,586 (9,094)
Hazard mitigation index (0-1)	242,496 (0.028) [21.58%]	88,420 (0.071) [8.00%]	118,294 **(0.108)** [11.45%] PROT: [11.3%]	56,384 **(0.086)** [5.19%]	1,040,962 (0.062)
Species richness (species)	**(653)** [52.3%]	(552) [6.5%]	(510) [5.3%] PROT: [24.7%]	(559) [3.5%]	(626)
Tree cover (M km^2) (%)	58.5 **(69.5)** [86.83%]	8.1 **(65.5)** [12.09%]	5.3 **(48.97)** [7.92%] PROT: [41.8%]	4.4 **(66.96)** [6.56%]	(40.15)
Protected area (M km^2) (%)	3.596 [72.4%]	0.45 [9.07%]	0.32 [6.48%]	0.25 [5.03%]	4.96
Deforestation (2000–2012)%	(2.59)	(2.85)	(1.78) PROT: (0.94)	(1.89)	(3.90)

[a]For areas with a positive water balance
[b]For areas with a positive water balance

Table 10.5 Ecosystem service provision by cloud forests, Africa

Ecosystem services	ALLF sum (mean)	MF sum (mean)	CAF sum (mean)	CF sum (mean)	ALL-L sum (mean)
Water (potential Mm^3 clean water produced)	2.8 (**0.84**) [34.8%]	1.38 (**0.74**) [17.1%]	0.5 (**0.74**) [6.4%] PROT: [12.8%]	0.3 (**0.9**) [3.3%]	8.1 (0.48)
Water (realised Mm^3 clean water produced)	2.79 (**0.84**) [34.8%]	1.38 (**0.75**) [17.2%]	0.51 (**0.74**) [6.37%] PROT: [12.7%]	0.26 (**0.88**) [3.28%]	8.03 (0.48)
AG carbon (M tonnes C storage) (tonnes C/km^2)	41003 (**12344**) [44.6%]	20854 (**11234**) [22.7%]	5680 (**8260**) [6.18%] PROT: [15.8%]	3896 (**13128**) [4.24%]	91790 (5471)
Soil carbon (M tonnes C storage) (tonnes C/km^2)	27857 (8386) [17.48%]	15687 (8451) [9.85%]	8422 (**12247**) [5.29%] PROT: [16.2%]	3913 (**13184**) [2.46%]	159303 (9495)
Carbon sequestration (M Dg/ha/yr) (Dg/ha/yr)	51.81 (**15.6**) [23.7%]	30.3 (**16.3**) [13.87%]	10.68 (**15.5**) [4.89%] PROT: [14.9%]	4.87 (**16.4**) [2.23%]	218 (13.01)
Hazard mitigation index (0-1)	0.002 (0.0006) [8.64%]	0.001 (0.0007) [5.6%]	0.0009 (0.001) [4.3%] PROT: [13.2%]	0.0002 (0.0006) [0.76%]	0.023 (0.001)
Species richness (species)	(475) [14.8%]	(499) [8.68%]	(553) [3.56%] PROT: [16.0%]	(543) [1.51%]	(636)
Tree cover (M km^2) (%)	214 (**64.6**) [56.9%]	116 (**62.8**) [30.9%]	28.6 (**41.5**) [7.58%] PROT: [16.0%]	20.2 (**68**) [5.35%]	377 (22.46)
Protected area (M km^2) (%)	0.51 [10.67%]	0.31 [6.51%]	0.1 [2.28%]	0.07 [1.53%]	4.73
Deforestation (2000–2012)%	(**2.12**)	(**2.02**)	(**1.95**) PROT: (0.60) [10.5%]	(**1.99**)	1.61

10 Mapping Hydrological Ecosystem Services and Impacts of Scenarios... 207

hazard mitigation and soil carbon storage services than ALLF. 0.1 of 0.58 Mkm^2 of CAF (17%) are protected leading to a two-third reduction of deforestation rates relative to unprotected CAFs but these protected areas secure relatively little of the CAF biodiversity, tree cover and ecosystem services (Table 10.5).

10.3.11 Tropical and Subtropical Asia

For tropical and subtropical Asia (Table 10.6) the mean per-unit-area clean water provision tends to be highest for ALLF and TMF (because of the relative lack of human influence on these forest types on this continent). For realised water provision all forest types have greater provision per-unit-area than ALL-L and that is also true for all carbon services (except soil carbon in MF, CAFs and CFs, which is similar to, or lower than, the mean for ALL-L). Hazard mitigation services are highest for CAFs. Species richness is greatest for CAFs, CFs and MFs whilst protected area coverage is significant for all forest types. ALLF represents only 29% of Asia's protected areas and CAFs a significant 17.7%. Deforestation rates are greater in ALLF. The rate in protected CAFs is 70% that of the all CAF rate so the protected areas do not seem very effective. The 20.5% of the study areas that are CAFs provide 22.9% of realised water, 28.9% of above-ground carbon, but 17.3% of soil carbon, 24.1% of hazard mitigation and 33% of the total tree cover. Their absolute contribution is significant relative to the ALLF class and for most services their provision per-unit-area is similar but less, though CAFs provide higher per-unit-area hazard mitigation services than ALLF. Some 0.1 of 3.43 Mkm^2 of CAF (13.9%) are protected leading to a one-third reduction of deforestation rates relative to unprotected CAFs, but these protected areas secure little (around 10%) of the total CAF biodiversity, tree cover and ecosystem services (Table 10.6).

10.3.12 Distance Decay of Hydrological Services

Though CAFs and CFs generally provide a higher density of ecosystem services than other forest types and in some continents have more local and downstream beneficiaries than other forest types, they remain relatively small in areal extent and thus their influence on national and continental natural capital and ecosystem service accounts will reflect this. The exception may be for water-related services where cloud forest types occupy important headwater locations with significant populations downstream. However, we cannot assume that all of these populations benefit equally from cloud forest hydrological services as the influence of a cloud forest (or any other hydrological unit) will decay with distance downstream. Figure 10.5 shows the hydrological footprint for CAFs in each continent, by pixel and country. It is clear from Fig. 10.5b, d, f and h that the influence of CAFs is highest in those CAFs that have an excess of local rainfall over evapotranspiration in all months and thus consistently produce runoff such that footprint can be 100% with such zones. As rivers leave CAFs, however, the CAF runoff is mixed with runoff

Table 10.6 Ecosystem service provision by different forest types, tropical and subtropical Asia

Ecosystem services	ALLF sum (mean)	MF sum (mean)	CAF sum (mean)	CF sum (mean)	ALL-L sum (mean)
Water (potential Mm^3 clean water produced)	7.81 **(1.35)** [57.6%]	2.97 **(1.12)** [21.9%]	3.06 **(0.89)** [22.6%] PROT: [10.9%]	2.27 **(1.08)** [16.7%]	13.5 (0.81)
Water (realised Mm^3 clean water produced)	7.64 **(1.33)** [57.6%]	2.94 **(1.12)** [22.2%]	3.03 **(0.88)** [22.9%] PROT: [10.6%]	2.25 **(1.08)** [16.9%]	13.3 (0.79)
AG carbon (M tonnes C storage) (tonnes C/km^2)	81066 **(14068)** [58.9%]	37441 **(14232)** [27.2%]	39683 **(11554)** [28.9%] PROT: [14.0%]	29079 **(13924)** [21.2%]	137441 (8192)
Soil carbon (M tonnes C storage) (tonnes C/km^2)	65966 **(11448)** [35.5%]	24148 (9179) [13.7%]	30404 (8852) [17.3%] PROT: [12.9%]	19246 (9215) [10.9%]	17602 (10491)
Carbon sequestration (M Dg/ha/yr) (Dg/ha/yr)	69.1 **(11.99)** [44.9%]	30.1 **(11.45)** [19.6%]	33.9 **(9.87)** [22.1%] PROT: [9.5 %]	23.1 **(11.1)** [15.1%]	153.7 (9.2)
Hazard mitigation index (0-1)	0.017 (0.0029) [35.68%]	0.008 (0.0028) [15.89%]	0.012 (0.0033) [24.1%] PROT: [11.0%]	0.006 (0.0028) [12.6%]	0.047 (0.0028)
Species richness (species)	(412) [32.8%]	**(441)** [16.1%]	**(435)** [20.7%] PROT: [10.1%]	**(443)** [12.8%]	(430)
Tree cover (M km^2) (%)	363 **(63.1)** [73.3%]	166 **(63.3)** [33.6%]	163 **(47.7)** [33.1%] PROT: [12.7%]	131 **(62.8)** [26.5%]	495 (29.5)
Protected area (M km^2) (%)	0.77 [28.8%]	0.45 [16.8%]	0.48 [17.7%]	0.35 [13.1%]	2.70
Deforestation (2000–2012) %	**(4.03)**	**(2.32)**	(1.83) PROT: (1.23) [8.8%]	(2.19)	(2.23)

from surrounding areas and so the influence of the CAFs decays quickly. A few rivers maintain influences of >50% throughout their course to the sea, others fall quickly to less than 10%. Transparent areas on the maps have a value close to zero. When averaged by country, it is clear that for certain small countries, cloud forests can influence up to 50% of the river flow. For small, wet countries such as Costa

10 Mapping Hydrological Ecosystem Services and Impacts of Scenarios... 209

Rica the contribution is much less. For larger countries the hydrological footprint of cloud forests is generally less than 20%, though it may be greater in lowland dry seasons (not shown) if CAFs continue to receive rainfall when the lowlands do not (Fig. 10.5).

If we calculate the hydrological footprint using WaterWorld outputs that include the peculiar hydrology of CAFs better than global rainfall and evapotranspiration datasets are able to, we see that many more of the cloud forests areas show all-year

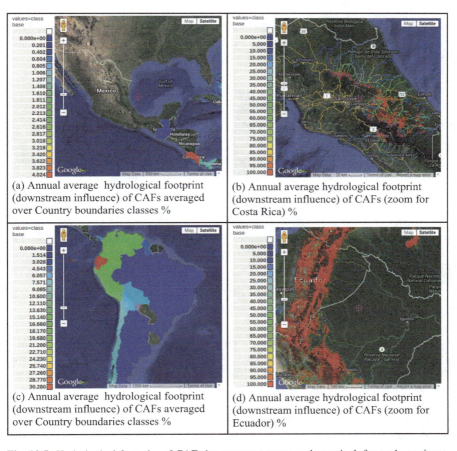

Fig. 10.5 Hydrological footprint of CAFs by country average and per pixel, for each continent. Background map data: Google, NASA, Terrametrics. (**a**) Annual average hydrological footprint (downstream influence) of CAFs averaged over country boundaries classes %, (**b**) annual average
Fig. 10.5 (continued) hydrological footprint (downstream influence) of CAFs (zoom for Costa Rica) %, (**c**) annual average hydrological footprint (downstream influence) of CAFs averaged over country boundaries classes %, (**d**) annual average hydrological footprint (downstream influence) of CAFs (zoom for Ecuador) %, (**e**) annual average hydrological footprint (downstream influence) of CAFs averaged over country boundaries classes %, (**f**) annual average hydrological footprint (downstream influence) of CAFs (zoom for Kenya) %, (**g**) annual average hydrological footprint (downstream influence) of CAFs averaged over country boundaries classes %, (**h**) annual average hydrological footprint (downstream influence) of CAFs (zoom for Bhutan) %

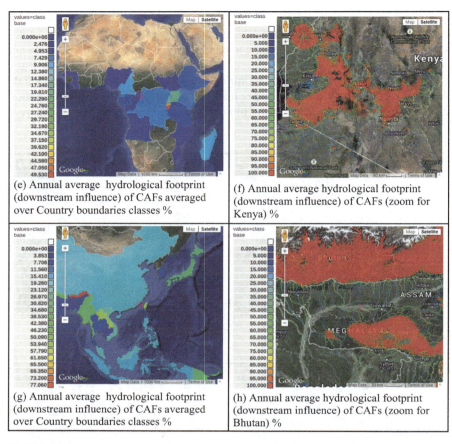

Fig 10.5 (continued)

rainfall excess over evapotranspiration and thus generate continuous runoff and a 100% local hydrological footprint. This inflates the national average for small, wet countries with a significant CAF area like Costa Rica (Fig. 10.6a) in this case from around 4% to 30%. For larger, drier countries like Madagascar the national average calculated with the WaterWorld data (8.5%) is similar to that calculated with the remote sensing data (9%). This gives some confidence that the remote-sensing based hydrological footprint results are reasonable pantropically even if they underestimate the footprint of CAFs in small, CAF dominated countries (Fig. 10.6).

Fig. 10.6 (continued) of CAFs averaged over country boundaries classes %, (**b**) annual average hydrological footprint (downstream influence) of CAFs %, (**c**) annual average hydrological footprint (downstream influence) of CAFs averaged over country boundaries classes %, (**d**) annual average hydrological footprint (downstream influence) of CAFs %, (**e**) annual % of runoff generated by fog averaged over major sub-basins (Hydrosheds) classes (%) and (**f**) annual % of runoff generated by fog averaged over major sub-basins (Hydrosheds) classes (%)

10 Mapping Hydrological Ecosystem Services and Impacts of Scenarios... 211

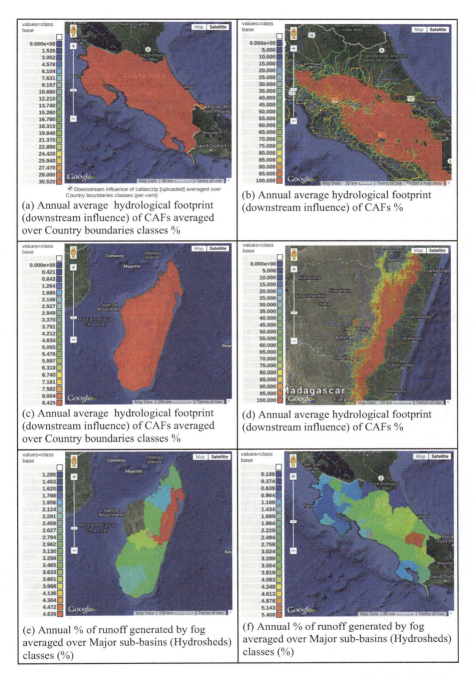

Fig. 10.6 Hydrological footprint of CAFs in Costa Rica and Madagascar, based on WaterWorld data and annual % of runoff generated by fog averaged over major sub-basins. Background map data: Google, NASA, Terrametrics: (**a**) annual average hydrological footprint (downstream influence)

Moreover, another of WaterWorld's outputs, the "annual percentage of runoff generated by fog" indicates the key additional flux of water associated with the cloud forest condition. These fog inputs are considered a key ecosystem service of CAFs (Bennett et al. 2009; Breuer et al. 2013; Bruijnzeel et al. 2011) but fog contributions to runoff decay quickly as distance downstream of cloud forest increases (Fig. 10.6e, f). This is because rainfall inputs from surrounding lands are much greater in magnitude and areal extent than fog inputs from CAFs, so that averaged over major basins fog contributions remain below 5%.

10.4 Discussion and Conclusions

10.4.1 Cloud Forests and their Beneficiaries

Cloud forests (both CAFs and CFs) are much more spatially restricted than lowland forests in all continents except Asia where CAFs are relatively extensive. Whilst forests represent nearly 30% of the tropical land area, CAFs and MFs represent only a third of these forests and CFs represent only 18%. These are significant areas pantropically but represent a small fraction of global forest cover, which is dominated by lowland tropical and boreal forests. Ecosystem services for all tropical forest types are relatively poorly known (Brandon 2014).

Only 3.6% of the population of the tropics inhabit CAFs, though densities are higher than for lowland forests in some continents, e.g. South America and higher than the continent average for Central America and Africa. However CAFs do occupy important headwater areas and have much larger downstream populations than other forest types, especially in Africa and Asia where each square kilometre of CAF has, on average, hundreds of thousands of downstream people. CAFs also have more downstream dams than any other forest type, with each square kilometre of CAF providing for an average of around 0.4 dams downstream, or put another way, one in two square km of CAF being upstream of a dam. Protected CAFs tend to be upstream of greater population densities and more dams than unprotected CAFs, but not exclusively so between the continents. However, downstream populations for protected CAFs vary from tens of thousands in Central and South American protected CAFs through to hundreds of thousands for Asian and African protected CAFs. These are a small fraction of the millions to billions of people in these continents that depend on ecosystem services from other systems (Summers et al. 2012).

10.4.2 The Ecosystem Services of Cloud Forests, Continent by Continent

In general, forested land tends to provide greater hydrological, carbon and hazard mitigation services than the continental averages as well as harbouring more species and, of course, tree cover. Different forest types have different concentrations of ecosystem services across the continents but in general CAFs and CFs tend to provide the highest densities of hydrological services with CFs also providing high densities of carbon storage, hazard mitigation and species richness in some continents. Though the concentration of this natural capital and ecosystem services is high in cloud forest types, their contribution to the total national or continental stocks and flow is low by virtue of their limited extent compared with other forest types.

There are also significant differences between continents. Figure 10.7a shows the mean for each variable within CAFs by continent. Since the variables have different units, they are expressed as a percentage of the maximum for that variable by

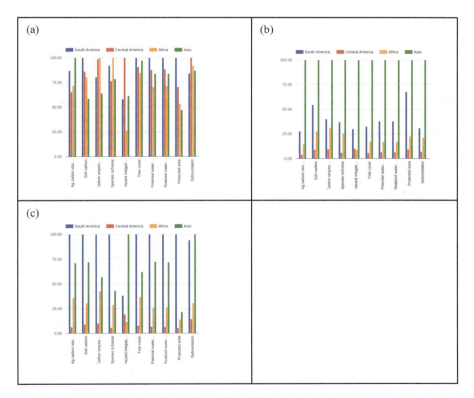

Fig. 10.7 (**a**) Mean (density) of variables within CAFs: comparison between continents and (**b**) total (sum) of variables within CAFs: comparison between continents, (**c**) total (sum) of variables within all forests: comparison between continents

continent. Clearly South America has the "richest" CAFs for many ecosystem services, followed by Africa (carbon sequestration and species richness) and Central America (hazard mitigation). Asia tends to have the lowest density of ecosystem service provision by CAFs. Mean protected area coverage of CAFs is highest in Latin America and lowest in Asia, whilst mean deforestation in CAFs is highest in Central America and lowest in South America. For the sum of these variables within CAFs (Fig. 10.7b) Asia is highest for all variables (having the greatest extent of CAFs), followed by South America, Africa and Central America. Though Asia has the lowest fraction of CAFs protected, it has the highest absolute area of protected CAFs, followed by South America and also the highest deforestation of CAFs, again followed by South America.

The situation is profoundly different when one examines all forest types (Fig. 10.7c) in which case the Amazon forests dominate most ecosystem service totals, bringing South America well ahead of Asia and Africa for all services except hazard mitigation and for species richness, despite the continent being smaller than Asia. Asia tops the continents for hazard mitigation services, reflecting the mountainous nature of the continent and importance of forests in mitigating those hazards. South America has the greatest protected forest area but also close to the greatest deforestation rate.

Clearly Asia is a very significant continent with respect to ecosystem services value of CAFs but also with respect to the need to better protect these rapidly disappearing forests. This is in contrast to the historic research effort on cloud forests with a Google Scholar search for ["cloud forest" central America] returning 17,200 results (1,270 including the additional term "ecosystem services"), ["cloud forest" south America]: 15,500 (1,140 with "ecosystem services"), ["cloud forest" Africa]: 9,480 (1,010 with "ecosystem services"), but ["cloud forest" Asia] only 6,590 (680 with "ecosystem services"). The body of research, including ecosystem services research, in cloud forests is highly skewed towards Central and South America, yet the most important cloud forests for most people locally and downstream are in Africa and Asia. The next decade should see a much greater focus on these important but pressured and rapidly changing forest regions.

10.4.3 The Implications of Distance Decay for Ecosystem Services from Cloud Forests

We have provided data on the ecosystem service contributions of tropical forest types by continent and forest type. Ecosystem service provision has been calculated on the basis of both potential and realised services. Some realised services such as carbon storage and sequestration have global beneficiaries meaning that proximity to the sites of service production is not necessary to receive the benefits. Others require beneficiaries to be downhill or downstream of the sites of service production. These services are much more challenging to account for because, whilst one can sum the number of potential beneficiaries (e.g. people, dams) downstream, it is not necessarily the case that all of these people are significantly affected by the cloud forest upstream. This is because there is a distance decay function for these services that is determined by the size of the site of production in relation to all other sites of production between the site of production and the ecosystem service consumer. We use the hydrological footprint and the fog contribution to runoff to indicate that, by nature of their often distant and spatially restricted nature, cloud forests rarely have more than 10% influence on the flow of rivers. This fundamental hydrological principle needs to be observed in analyses of the influences of cloud forests and other spatially restricted areas (such as protected areas) on ecosystem services downstream.

By calculating the downstream difference in hydrological footprint from each pixel to its downstream neighbour and then masking all but the negative values (i.e. downstream decay) we can examine the mean decay in hydrological influence per cell (sq km) (Fig. 10.8). This is highest at the CAF, non-CAF boundary of course but also occurs along the full length of rivers exiting CAFs. Values for Asian countries (Fig. 10.8) vary from around −0.3%/km to −25%/km, indicating a decline to imperceptible influence within around 333–7 km, respectively, downstream, depending on local conditions. These values are dominated by the higher CAF-non CAF boundary decay values, with values within river systems being much smaller. For the "annual percent of runoff generated by fog" variable for Madagascar, the rate of

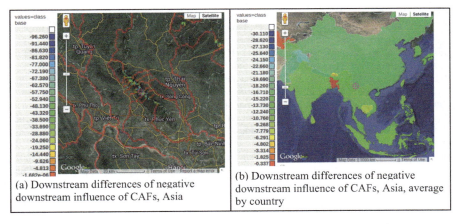

Fig. 10.8 Downstream decay of hydrological footprint: example from Asia. Background map data: Google, NASA, Terrametrics: (**a**) downstream differences of negative downstream influence of CAFs, Asia and (**b**) downstream differences of negative downstream influence of CAFs, Asia, average by country

decay downstream is -0.88% within the CAF area and -0.3% overall, indicating decay to indiscernible within 133–333 km downstream. The "half-life" of a cloud forest downstream hydrological influence can thus be considered as 166 km with an influence drop of 25–50% at the CAF, non-CAF boundary and then a fraction of 1% per km downstream thereafter. However, there are no rules of thumb. Because of the significant site-to-site variability and thus range of downstream decay functions, these values should be calculated on a site-by-site basis and can be so with WaterWorld. They are fundamental to understanding the relationship between production and consumption of hydrological ecosystem services, especially for remote, isolated systems like cloud forests.

Acknowledgements The author gratefully acknowledges all the institutions that make global datasets available for the scientific community to use.

References

Bennett EM, Peterson GD, Gordon LJ (2009) Understanding relationships among multiple ecosystem services. Ecol Lett 12:1394–1404

BIRDLIFE (2012) Birdlife International IUCN red list for birds. http://www.iucnredlist.org/technical-documents/spatial-data#birds

Brandon K (2014) Ecosystem services from tropical forests: review of current science. CGD working paper 380. Washington, DC: Center for Global Development. http://www.cgdev.org/publication/ecosystem-services-tropical-forests-review-currentscience-working-paper-380

Brauman KA, Daily GC, Duarte TKE, Mooney HA (2007) The nature and value of ecosystem services: an overview highlighting hydrologic services. Annu Rev Environ Resour 32:67–98

Breuer L, Windhorst D, Fries A, Wilcke W (2013) Supporting, regulating, and provisioning hydrological services. In: Ecosystem services, biodiversity and environmental change in a tropical mountain ecosystem of South Ecuador. Springer, Berlin, pp 107–116

Bruijnzeel LA, Mulligan M, Scatena FS (2011) Hydrometeorology of tropical montane cloud forests: emerging patterns. Hydrol Process 25:465–498

Bubb P, May I, Miles L, Sayer J (2004) Cloud forest agenda. UNEP-World Conservation Monitoring Centre, Cambridge, UK. http://sea.unep-wcmc.org/forest/cloudforest/index.cfm

Center For International Earth Science Information Network (CIESIN) (2002) Country-level GDP and downscaled projections based on the A1 A2 B1 and B2 Marker Scenarios 1990-2100. CIESIN Columbia University, Palisades, NY. http://www.ciesin.columbia.edu/datasets/downscaled

Dapples F, Lotter AF, Van Leeuwen JF, Van Der Knap WO, Dimitriadis S, Oswald D (2002) Paleolimnological evidence for increased landslide activity due to forest clearing and land-use since 3600 cal BP in the western Swiss Alps. J Paleolimnol 27:239–248

De Groot RS, Alkemade R, Braat L, Hein L, Willemen L (2010) Challenges in integrating the concept of ecosystem services and values in landscape planning, management and decision making. Ecol Complex 7:260–272

Dilley M (2005) Natural disaster hotspots: a global risk analysis. Version 1.0. Disaster risk management series, No. 5. World Bank, Washington DC. https://openknowledge.worldbank.org/handle/10986/7376

FAO (1998) FRA 2000. Terms and definitions. FRA working paper 1, FAO Forestry Department. http://www.fao.org/forestry/fo/fra/index.jsp. Under publications

FAO (2010) GIEWS: world road trails. Whole world's roads and railways. http://ldvapp07.fao. org:8030/downloads/layers/world_roadstrail.xml

Farr TG, Kobrick M (2000) Shuttle radar topography mission produces a wealth of data. Eos Trans 81:583–585

Fisher B, Turner RK, Morling P (2009) Defining and classifying ecosystem services for decision making. Ecol Econ 68:643–653

Funnell DC, Price MF (2003) Mountain geography: a review. Geogr. J 169:183–190

Hansen M, Defries R, Townsend JR, Carroll M, Dimiceli C, Sohlberg R (2006) Vegetation continuous fields MOD44B, 2001 percent tree cover, collection 4. University of Maryland, College Park, MDaryland, 2001. http://glcf.umd.edu/data/vcf/

Hijmans RJ, Cameron SE, Para JL, Jones PG, Jarvis A (2005) Very high resolution interpolated climate surfaces for global land areas. Int J Climatol 25:1965–1978

Ho LC (1976) Variation in the carbon/dry matter ratio in plant material. Ann Bot 40:163–165

IUCN (2010) An analysis of reptiles on the 2010 IUCN red list. http://www.iucnredlist.org/technical-documents/spatial-data

IUCN, Conservation International, and Nature Reserve (2008) An analysis of amphibians on the 2008 IUCN red list. http://www.iucnredlist.org/amphibians

IUCN, Conservation International, Arizona State University, Texas A&M University, University of Rome, University OF Virginia, Zoological Society London (2008) An analysis of mammals on the 2008 IUCN Red List. http://www.iucnredlist.org/mammals

Jarvis A, Mulligan M (2009) The climate of cloud forests. Hydrol Process 25:327–343

Jenkins M, Scherr SJ, Inbar M (2004) Markets for biodiversity services: potential roles and challenges. Environ Sci Policy Sustain Dev 46:32–42

Jones R (2002) Algorithms for using a DEM for mapping catchment areas of stream sediment samples. Comput Geosci 28:1051–1060

LandScan™ (2007) LandScan™ global population database. Oak Ridge National Laboratory, Oak Ridge, TN. http://www.ornl.gov/landscan/

Lehner B, Doll P (2004) Development and validation of a global database of lakes, reservoirs and wetlands. J Hydrol 296:1–22

Lehner B, Verdin K, Jarvis A (2008) New global hydrography derived from spaceborne elevation data. Eos Trans AGU 89:93–94

Mu Q, Zhao M, Running SW (2011) Improvements to a MODIS global terrestrial evapotranspiration algorithm. Remote Sens Environ 115:1781–1800

Mulligan M (2009a) Global mean dry matter productivity based on SPOT-VGT (1998–2008). http://geodata.policysupport.org/dmp

Mulligan M (2009b) The human water quality footprint: agricultural, industrial, and urban impacts on the quality of available water globally and in the Andean region. In: Proceedings of the international conference on integrated water resource management and climate change, Cali, Colombia, p 11.

Mulligan M (2010a) Modeling the tropics-wide extent and distribution of cloud forest and cloud forest loss, with implications for conservation priority. In: Bruijnzeel LA, Scatena FN, Hamilton LS (eds) Tropical Montane Cloud forests: science for conservation and management. Cambridge University Press, Cambridge, pp 14–39

Mulligan M (2010b) A combined global database of mines. http://geodata.policysupport.org/mines

Mulligan M (2010c) A combined global database of oil and gas wells. http://geodata.policysupport.org/oilandgas

Mulligan M (2010d) SimTerra: A consistent global gridded database of environmental properties for spatial modelling. http://www.policysupport.org/simterra

Mulligan M (2013) WaterWorld: a self-parameterising, physically-based model for application in data-poor but problem-rich environments globally. Hydrol Res 44(5):748–769

Mulligan M (2015a) Climate change and food-water supply from Africa's drylands: local impacts and teleconnections through global commodity flows. Int J Water Resour Dev 31:450–460

Mulligan M (2015b) Trading off agriculture with nature's other benefits, spatially. In: Zolin CA, De Rodrigues RAR (eds) Impact of climate change on water resources in agriculture. CRC Press, Boca Raton, FL, pp 184–205

Mulligan M, Clifford NA (2015) Is managing ecosystem services necessary and sufficient to ensure sustainable development? In: Springet D, Redclift M (eds) Handbook of sustainable development. Routledge, Abingdon, UK, pp 179–195

Mulligan MA, Guerry K, Arkema K, Bagstad D, Villa F (2010) Capturing and quantifying the flow of ecosystem services. In: Silvestri S, Kershaw F (eds) Framing the flow: innovative approaches to understand, protect and value ecosystem services across linked habitats. UNEP World Conservation Monitoring Centre, Cambridge, UK

Mulligan M, Saenz L, Van Soesbergen A (2011) Development and validation of a georeferenced global database of dams. http://geodata.policysupport.org/dams

Naidoo R, Balmford A, Costanza R, Fisher B, Green RE, Lehner B, Malcolm TR, Ricketts TH (2008) Global mapping of ecosystem services and conservation priorities. Proc Natl Acad Sci 105:9495–9500

National Geophysical Data Center/World Data Center (NGDC/WDC) (2011) Historical tsunami database, Boulder, CO, USA. http://www.ngdc.noaa.gov/hazard/tsu_db.shtml

Pena-Arancibia JL, Van Dijk AI, Guerschman JP, Mulligan M, Bruijzeel LA, McVicar TR (2012) Detecting changes in streamflow after partial woodland clearing in two large catchments in the seasonal tropics. J Hydrol 416:60–71

Ramankutty N, Evan AT, Monfreda C, Foley JA (2008) Farming the planet: 1. Geographic distribution of global agricultural lands in the year 2000. Glob Biogeochem Cycles 22:1–19

Ruesch A, Gibbs HK (2008) New IPCC tier-1 global biomass carbon map for the year 2000. Available online from the Carbon Dioxide Information Analysis Center, Oak Ridge National Laboratory, Oak Ridge, Tennessee. http://cdiac.ornl.gov

Saatchi S, Harris NL, Brown S, Lefsky M, Mitchard ET, Salas W, Zutta BR, Buermann W, Lewis SL, Hagen S, Petrova S, White L, Silman M, Morel A (2011) Benchmark map of forest carbon stocks in tropical regions across three continents. Proc Natl Acad Sci 14:9899–9904

Scharlemann JPW, Hiederer R, Kapos V (2009) Global map of terrestrial soil organic carbon stocks. A 1-km dataset derived from the harmonized World soil database. UNEP-WCMC & EU-JRC, Cambridge

Schneider A, Friedl MA, Potere D (2009) A new map of global urban extent from MODIS data. Environ Res Lett 4:1–11

Siebert S, Doll P, Feick S, Fremken K, Hoogeveen J (2007) Global map of irrigation areas version 4.0.1. University of Frankfurt (Main), Germany, and FAO, Rome, Italy. http://www.fao.org/nr/water/aquastat/irrigationmap/index10.stm

Spalding MD, Blasco F, Fields CD (eds) (1997) World Mangrove Atlas. The International Society for Mangrove Ecosystems, Okinawa, p 178

Summers JK, Smith LM, Case JL, Linthurst RA (2012) A review of the elements of human well-being with an emphasis on the contribution of ecosystem services. Ambio 41:327–340

Townshend JRG, Carroll M, Dimiceli C, Sohlberg R, Hansen M, Defries R (2011) Vegetation continuous fields MOD44B, 2001 percent tree cover, collection 5. University of Maryland, College Park, Maryland. http://glcf.umd.edu/data/vcf/

USGS (2006) Shuttle radar topography mission water body dataset. http://edc.usgs.gov/products/elevation/swbd.html

World Database on Protected Areas (WDPA) (2014) Annual release 2014. www.protectedplanet.net

Chapter 11
Conclusions, Synthesis, and Future Directions

Randall W. Myster

11.1 Conclusions

Like the Amazon the Andes is more than just mountains, grasslands, deserts, and forests. Its cloud forests in particular (Fig. 11.1) provide important ecosystem services for South America and help define various peoples and their cultures. In Chap. 1 I presented those cloud forests as occurring across large and important elevational and latitudinal gradients, where the edges of those gradients may be stressed by climatic conditions (see Fig. 11.2 as an artist's presentation). Physical effects of stress include low temperature, wind, and relatively little precipitation. Chemical effects of stress include lack of nutrients and chemical effects of burning. I also presented my research into cloud forests structure, function, and dynamics and the difficulties of cloud forest recruitment. In particular, the seed rain is probably not limiting cloud forest recruitment, regeneration, and plant community dynamics. What happens after dispersal is more limiting and predation took most of those seeds, more seeds than was taken by pathogens or that germinated.

The book started with a discussion of the tree line between cloud forest and paramó grassland suggesting that it is determined not just by climatic (temperature, precipitation) and local conditions (wind) but also human impacts, there was an ancient anthropogenic legacy of the different ways that people used tropical montane cloud forests landscapes in the past. Indeed, Andean treeline ecotone regions are cultural/socioecological landscapes that require more than traditional conservation approaches. Next, litterfall in the Andean cloud forest as a proxy for the total productivity of forests and a major vector of nutrient cycling was discussed. These forests become increasingly nutrient efficient with increasing elevation, while there is no indication of a general change in the kind of nutrient limitation. There may be systematic relationships between abiotic conditions and litterfall, which could be

R. W. Myster (✉)
Department of Biology, Oklahoma State University,
Oklahoma City, OK, USA

© Springer Nature Switzerland AG 2021
R. W. Myster (ed.), *The Andean Cloud Forest*,
https://doi.org/10.1007/978-3-030-57344-7_11

Fig. 11.1 The Andean cloud forest

Fig. 11.2 Wald Bau (forest-construction), 1919, by Paul Klee (1879–1940), mixed media chalk, 27 × 25 cm—Museo del Novecento, Milan, Italy

used to predict litterfall in the Andes. The observed elevational influence of litterfall in the humid tropical Andes suggests that the forest productivity will likely respond to climate change driving the vegetation belts to higher elevation with an unknown overall effect on C sequestration of these forests.

We saw that just about all investigated trees were arbuscular mycorrhizal (AM), each tree individual has an individual composition of AM fungal partners, which usually consists of one (two) common fungi and 3–8 rarely occurring fungi. The common species are not specialists, but generalists, from 1000 masl to 4000 masl, there is a high turnover in terms of both the AM fungi and the plant species. Only three species of fungi occur at all altitudes. Also, information gathered for the Tucuman Amazon parrot (*Amazona tucumana*) for 15 years show it to have high

11 Conclusions, Synthesis, and Future Directions

rates of nesting success, large clutches, and many fledglings per laying female. Nesting and spatial requirements could limit management, and so to ensure the conservation of this parrot outside protected areas managers should promote the retention of large *B. salicifolius*, *C. lilloi*, and *J. australis* birds and their nests that are selected for.

Cloud forests of the Andes are diverse and have a high degree of endemism, attributed to a mixture of historical, evolutionary, and ecological processes. For example, the relationship between species diversity and spatial structure of the Andean mountains, or the role of physical barriers that arose during the late Eocene (uplift of the Andes) and Neogene and the range shifts is caused by climate during the quaternary period. In addition, species tolerances to environmental conditions (e.g., temperature) can affect species dispersal and facilitate allopatric isolation. And finally, ecosystem services are the benefits that human populations derive from nature, here from cloud forests which occur in some important headwater areas with very significant human populations locally and downstream. We need to look at the benefits which accrue from these services, who benefits, and which of these benefits are protected for the future, within the context of deforestation trends.

11.2 Synthesis

Within the biology of any terrestrial ecosystem, both abiotic and biotic processes move, primarily, in and out of the plant phytomass (the biomass [Myster 2003] and necromass together). In this volume, and in my first four books, I have presented a view of terrestrial ecosystems as plant-centered, where components of ecosystems cycle in and out of—or flow through, like energy—the total plant phytomass. No other component or components of the Andean cloud forest ecosystem, except the phytomass, can assume this central role as a conduit for physical, chemical, and biological parts of the ecosystem (Myster 2001). Only the phytomass mediates and integrates between biogeochemical cycles (including cycles of productivity and decomposition: Myster 2003), conducting most an ecosystem's energy and nutrient processing. The phytomass should, therefore, be put in the center in our conceptual models of terrestrial ecosystems.

Ecosystem structure, function, and dynamics in the Andean Cloud forest is then fundamentally about the dynamics of the phytomass (Myster 2001, 2003) where biomass loss opens a patch of space where resources may be available (Bazzaz 1996). Then because plants are sessile organisms—and thus "seek" space—that patch of space can be occupied by other plants either by them growing into it themselves and/or by new plants recruiting into it. Ranking disturbances, using phytomass loss as the metric, creates a hierarchy which includes phytomass removed without whole plant death ("none" where no plant dies but some of its biomass was removed and/or some of its necromass decayed), with one plant dying ("one"), and with more than one plant dying as a group ("many"). And so, disturbances occur at different levels of this hierarchy of biomass loss where each disturbance at each level creates a discrete patch of space where plant responses may be present.

Those responses are conceptualized as plant-plant replacements (Myster 2018) from "none" where no new plant recruits into that now available patch of space but neighboring plants may grow into it, to "one" where one new plant recruits into that now available patch of space and may grow to occupy additional space in that patch, and/or to "many" where more than one plant as a group recruits into that now available patch of space and may grow to occupy additional space in that patch. For example, during old-field succession plants are often replaced by plants with the same size or larger seeds (Myster 2007). I posit that while natural disturbances may lead to plant-plant replacements like none => none, none => one, none => many, for example: Myster 2018), human-caused disturbances—because they remove relatively larger amounts of biomass—may lead to replacements like many => none, many => one, many => many. And so, the plant-plant replacements in an area could indicate the level and kind of past disturbance, just like crops and/or associated species still growing in an agricultural field after abandonment indicate the crop that was growing there before abandonment. There may also be a tendency for responses to match their disturbances (parallelism) that is none => none, one => one and many => many replacements are most common. These neighborhoods spaces may overlap: zero degree no overlap/open space, 1° two trees overlap, 2° three trees overlap etc.... And the more overlap the less likely other plants can use the space.

In conclusion, (1) biomass loss is the best way to measure disturbance (seen as severity in the disturbance regime) and to compare between disturbances (in this review we saw that severity differed fundamentally among all disturbances of the Andean cloud forest), (2) biomass can be lost as part of a plant, as the whole plant, or as many plants lost as a group, (3) disturbances can thus be placed on a hierarchy based on the relative amount of biomass lost, (4) at each point on a hierarchy, a patch of space is created as biomass is lost, and resources may become available within that patch of space, (5) living plants can then respond in that newly created patch of space, and (6) those responses are among nine classes of plant-plant replacement. These plant-plant replacements were documented among the responses and disturbances of the Andean cloud forest research papers in this review.

11.3 Future Directions

Chapters suggest the continued sampling of all ecosystem components in large Andean cloud forest plots with an emphasis on exploration of interactive links among ecosystem components. This should be followed by field experiments that need to be designed to find these links and should have a special focus on the early parts of regeneration (i.e. recruitment of seeds and saplings).

References

Bazzaz FA (1996) Plants in changing environments: linking physiological, population, and community ecology. Cambridge University Press, Cambridge

Myster RW (2001) What is ecosystem structure? Caribb J Sci 37:132–134

Myster RW (2003) Using biomass to model disturbance. Community Ecol 4:101–105

Myster RW (2007) Post-agricultural succession in the Neotropics. Springer-Verlag, Berlin

Myster RW (2018) The nine classes of plant-plant replacement. Ideas Ecol Evol 11:29–34

Correction to: Análisis Regional En Ecosistemas De Montaña En Colombia: Una mirada desde la funcionalidad del paisaje y los servicios ecosistémicos

Paola Isaacs-Cubides, Julián Díaz, and Tobias Leyva-Pinto

Correction to:
Chapter 3 in: R. W. Myster (ed.), *The Andean Cloud Forest,*
https://doi.org/10.1007/978-3-030-57344-7_3

This book was inadvertently published without the updated Figures 3.3, 3.5, 3.9, 3.10, 3.11, 3.12 and 3.13 (mentioned below) in chapter 3. This is now updated and corrected.

The updated online version of this chapter can be found at
https://doi.org/10.1007/978-3-030-57344-7_3

© Springer Nature Switzerland AG 2021
R. W. Myster (ed.), *The Andean Cloud Forest,*
https://doi.org/10.1007/978-3-030-57344-7_12

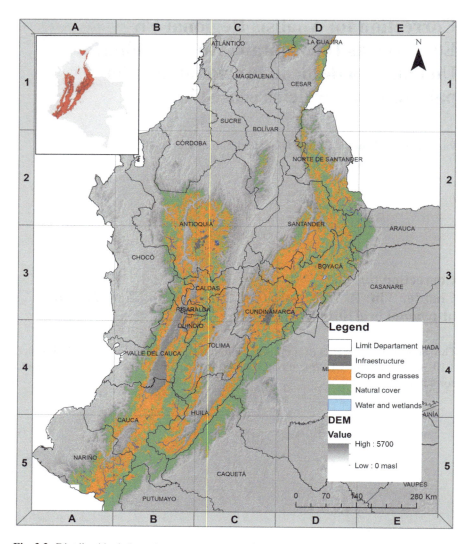

Fig. 3.3 Distribución de las coberturas presentes en las zonas de montaña

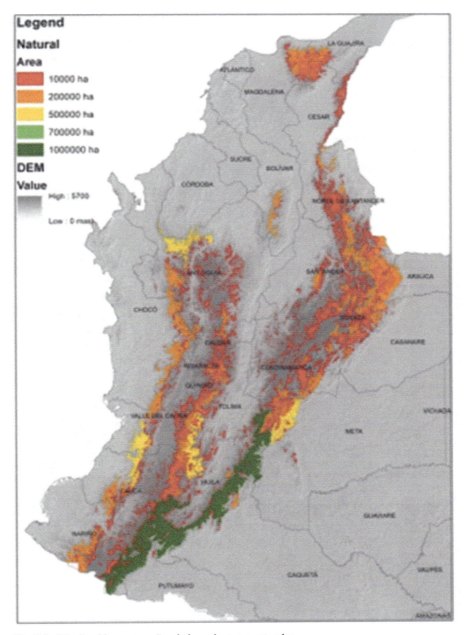

Fig. 3.5 Distribución por tamaños de las coberturas naturales

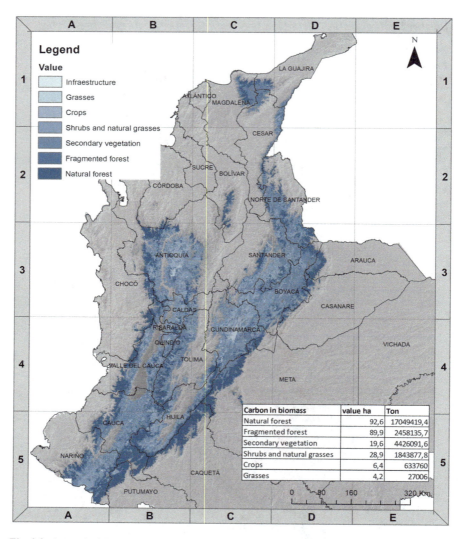

Fig. 3.9 Acumulación de carbono en biomasa, de acuerdo a los tipos de cobertura

Correction to: Análisis Regional En Ecosistemas De Montaña En Colombia

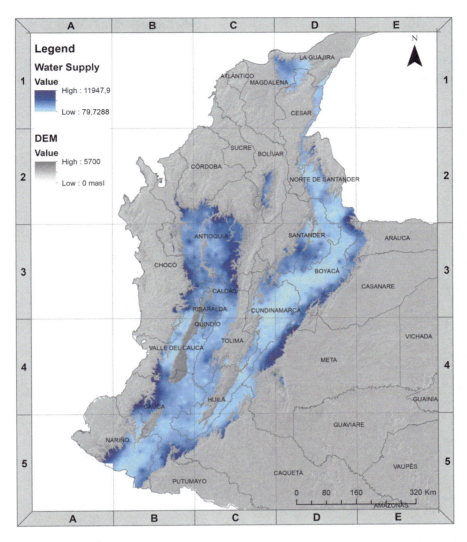

Fig. 3.10 Oferta hídrica calculada para las zonas de montaña

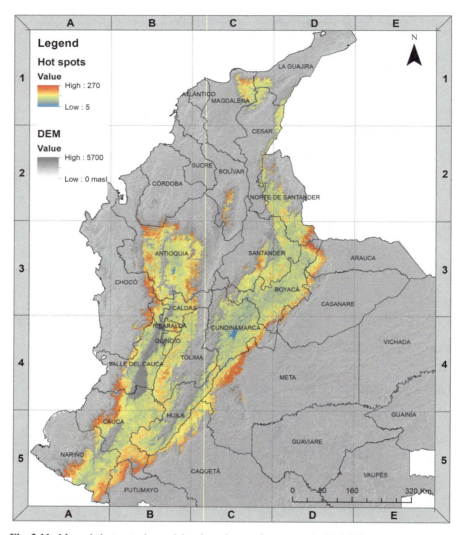

Fig. 3.11 Mapa de hotspots de servicios de carbono, oferta y regulación hídrica

Correction to: Análisis Regional En Ecosistemas De Montaña En Colombia C7

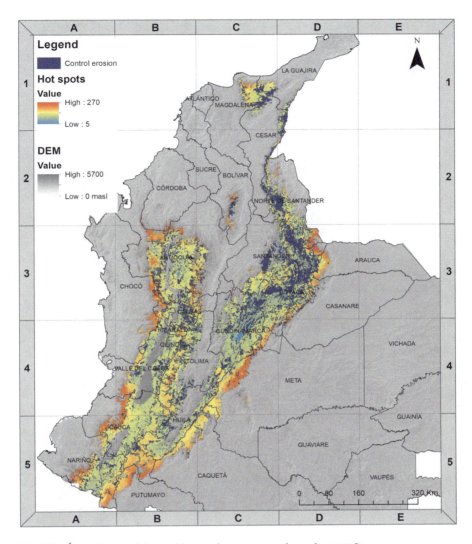

Fig. 3.12 Áreas de control de erosión en coberturas vegetales en la montaña

Correction to: Análisis Regional En Ecosistemas De Montaña En Colombia

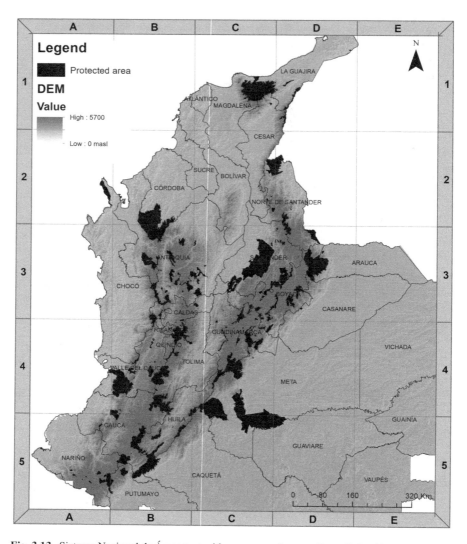

Fig. 3.13 Sistema Nacional de Áreas protegidas en zonas de montaña en Colombia

Printed in the United States
By Bookmasters